VOLUME ONE HUNDRED AND FORTY NINE

PROGRESS IN MOLECULAR BIOLOGY AND TRANSLATIONAL SCIENCE

Gene Editing in Plants

VOLUME ONE HUNDRED AND FORTY NINE

PROGRESS IN MOLECULAR BIOLOGY AND TRANSLATIONAL SCIENCE

Gene Editing in Plants

Edited by

DONALD P. WEEKS

Department of Biochemistry,
University of Nebraska,
Lincoln, NE, United States

BING YANG

Department of Genetics,
Development and Cell Biology,
Iowa State University,
Ames, IA, United States

ELSEVIER

ACADEMIC PRESS

An imprint of Elsevier

Academic Press is an imprint of Elsevier
525 B Street, Suite 1800, San Diego, CA 92101-4495, United States
50 Hampshire Street, 5th Floor, Cambridge, MA 02139, United States
125 London Wall, London EC2Y 5AS, United Kingdom
The Boulevard, Langford Lane, Kidlington, Oxford OX5 1GB, United Kingdom

First edition 2017

ISBN: 978-0-12-811743-9
ISSN: 1877-1173

Publisher: Zoe Kruze
Acquisition Editor: Alex White
Editorial Project Manager: Helene Kabes
Production Project Manager: James Selvam
Cover Designer: Matthew Limbert

Typeset by SPi Global, India

CONTENTS

CONTRIBUTORS

Nicholas J. Baltes
Calyxt, Inc., New Brighton, MN, United States

Honghao Bi
Iowa State University, Ames, IA, United States

Doane Chilcoat
DuPont Pioneer, Johnston, IA, United States

Xinhua Dai
Section of Cell and Developmental Biology, University of California San Diego, La Jolla, CA, United States

John P. Davies
Dow AgroSciences, Indianapolis, IN, United States

Deqiang Duanmu
State Key Laboratory of Agricultural Microbiology, College of Life Science and Technology, Huazhong Agricultural University, Wuhan, China

Javier Gil-Humanes
Calyxt, Inc., New Brighton, MN, United States

Yubing He
National Key Laboratory of Crop Genetic Improvement and National Center of Plant Gene Research (Wuhan), Huazhong Agricultural University, Wuhan, China

Sandeep Kumar
Dow AgroSciences, Indianapolis, IN, United States

Yao-Guang Liu
State Key Laboratory for Conservation and Utilization of Subtropical Agro-Bioresources; College of Life Sciences, South China Agricultural University, Guangzhou, China

Zhan-Bin Liu
DuPont Pioneer, Johnston, IA, United States

Xingliang Ma
Global Institute for Food Security, University of Saskatchewan, Saskatoon, SK, Canada

Magdy M. Mahfouz
Laboratory for Genome Engineering, 4700 King Abdullah University of Science and Technology, Thuwal, Saudi Arabia

Bastian Minkenberg
The Pennsylvania State University, University Park, PA, United States

Keishi Osakabe
Faculty of Bioscience and Bioindustry, Tokushima University, Tokushima, Japan

Yuriko Osakabe
Faculty of Bioscience and Bioindustry, Tokushima University, Tokushima, Japan

Jeffry Sander
DuPont Pioneer, Johnston, IA, United States

Lakshmi Sastry-Dent
Dow AgroSciences, Indianapolis, IN, United States

Manal Tashkandi
Laboratory for Genome Engineering, 4700 King Abdullah University of Science and Technology, Thuwal, Saudi Arabia

Daniel F. Voytas
Calyxt, Inc., New Brighton, MN, United States

Longlong Wang
State Key Laboratory of Agricultural Microbiology, College of Life Science and Technology, Huazhong Agricultural University, Wuhan, China

Longxiang Wang
State Key Laboratory of Agricultural Microbiology, College of Life Science and Technology, Huazhong Agricultural University, Wuhan, China

Rongchen Wang
National Key Laboratory of Crop Genetic Improvement and National Center of Plant Gene Research (Wuhan), Huazhong Agricultural University, Wuhan, China

Donald P. Weeks
University of Nebraska, Lincoln, NE, United States

Matthew Wheatley
The Pennsylvania State University, University Park, PA, United States

Jeffrey D. Wolt
Biosafety Institute for Genetically Modified Agricultural Products; Crop Bioengineering Consortium, Iowa State University, Ames, IA, United States

Xianrong Xie
State Key Laboratory for Conservation and Utilization of Subtropical Agro-Bioresources; College of Life Sciences, South China Agricultural University, Guangzhou, China

Bing Yang
Iowa State University, Ames, IA, United States

Yinong Yang
The Pennsylvania State University, University Park, PA, United States

Syed Shan-e-Ali Zaidi
Laboratory for Genome Engineering, 4700 King Abdullah University of Science and Technology, Thuwal, Saudi Arabia; National Institute for Biotechnology and Genetic Engineering (NIBGE), Faisalabad, Pakistan

Yaling Zhang
State Key Laboratory for Conservation and Utilization of Subtropical Agro-Bioresources; College of Life Sciences, South China Agricultural University, Guangzhou, China

Yunde Zhao
Section of Cell and Developmental Biology, University of California San Diego, La Jolla, CA, United States

Yu Zhou
State Key Laboratory of Agricultural Microbiology, College of Life Science and Technology, Huazhong Agricultural University, Wuhan, China

PREFACE

Rapid development of new techniques for gene and genome editing is catalyzing phenomenal progress in basic biological research with animal, plant, and microbial systems as well as fostering practical applications in medicine, industry, and agriculture. For laboratory scientists using animals and plants for their research, there had been until recently a great envy of investigators who were using organisms (e.g., bacterial and yeast) whose genes could be precisely modified with relative ease in the laboratory. Such genetic modifications allowed a multitude of fundamental discoveries at a remarkable pace during the second half of the 20th century. These discoveries greatly expanded the knowledge of basic cell and molecular biology and also gave rise to incredible advances in many applied fields of science—including birth of the biotechnology industry. The advent of facile, but expensive, methods for inactivating or modifying specific genes in the mouse paved the way for important new understandings of mammalian development, physiology, and diseases—and, with this new knowledge, the development of improvements in medical care for humans and other animals. With the invention of zinc finger nuclease technology (and, later meganuclease technology) came the first break for researchers who were using organisms heretofore recalcitrant to gene-editing procedures. With these new tools, scientists could embark on new avenues of investigations that formerly were confined to use with organisms that could easily insert exogenous DNA sequences into their genomes using homologous recombination. Subsequent development of TALEN and CRISPR/Cas9 technologies, with their greatly improved ease of use and targeting specificities, has made available the power of gene editing to a broad spectrum of life scientists around the world.

In this book, we focus on exciting new developments in gene-editing technologies for plant biology research and their applications to plant-based production in agriculture. Chapter 1 by Baltes, Gil-Humanes, and Voytas provides an overview of the development of genome editing for use in agriculture and the opportunities and challenges that lie ahead. Zhang, Ma, Xie, and Liu (Chapter 8) provide an overview of the various gene-editing techniques available for plant biologists and recommendations regarding their use, while Davies, Kumar, and Sastry-Dent (Chapter 3) provide an in-depth discussion of zinc finger nucleases and their uses in crop improvement. Other chapters describe improvements in gene-editing techniques such as multiplex gene editing using either tRNA-processing systems

(Chapter 7 by Minkenberg, Wheatley, and Yang) or ribozyme-processing systems (Chapter 9 by He, Wang, Dai, and Zhao) for sgRNA production during multiplex gene targeting. Application of gene editing to improving plant response to abiotic stresses is addressed by Osakabe and Osakabe (Chapter 6), while efforts to develop plants with improved immunity to plant viruses are outlined by Zaidi, Tashkandi, and Mahfouz (Chapter 10). Wang, Wang, Zhou, and Duanmu describe how the CRISPR/Cas9 system can be effectively used to explore the complex systems associated with nitrogen fixation in legumes (Chapter 11). A few of the chapters focus on specific crops including maize and soybeans (Chapter 2 by Chilcoat, Liu, and Sander), rice (Chapter 5 by Bi and Yang), and the broad spectrum of polyploid plants that account for a large portion of crop production worldwide (Chapter 4 by Weeks). Finally, Wolt (Chapter 12) reviews important issues surrounding the governance and regulation of gene-edited plants and how these issues will play a pivotal role in how the technology will (or will not) benefit consumers in countries around the world.

It has been our goal in assembling this book that it will serve as a valuable source of information (and potentially inspiration) to students, scientists, and others who wish to learn about recent advances in gene-editing technologies and/or are contemplating the use of these technologies in their professional lives. We believe these technologies will no doubt revolutionize research in the plant sciences and have the potential to contribute significantly to the challenging task of feeding the rapidly increasing (and growingly affluent) world population.

We wish to acknowledge Dr. P. Michael Conn for his initial invitation to us to prepare this book and his steadfast encouragement along the way. The *Progress in Molecular Biology and Translational Science* series will serve as a concrete and continuing testament to his significant contributions to science. We sincerely thank Senior Editorial Project Manager, Ms. Helene Kabes, for her highly professional assistance during the preparation of this book and for her dedication to making the project a success. We and other authors of chapters in this book are indebted to Project Manager, Mr. James Selvam, for his efficient and conscientious final assembly and editing of our contributions. Finally, we would like to thank the outstanding authors of chapters in this book who volunteered their time and strong efforts to produce well-written and well-documented contributions that reflect the state-of-the-art science in gene editing in plants.

Donald P. Weeks
University of Nebraska, Lincoln, NE, United States
Bing Yang
Iowa State University, Ames, IA, United States

CHAPTER ONE

Genome Engineering and Agriculture: Opportunities and Challenges

Nicholas J. Baltes[1], Javier Gil-Humanes, Daniel F. Voytas
Calyxt, Inc., New Brighton, MN, United States
[1]Corresponding author: e-mail address: nbaltes@calyxt.com

Contents

Progress in Molecular Biology and Translational Science, Volume 149
ISSN 1877-1173
http://dx.doi.org/10.1016/bs.pmbts.2017.03.011

Abstract

In recent years, plant biotechnology has witnessed unprecedented technological change. Advances in high-throughput sequencing technologies have provided insight into the location and structure of functional elements within plant DNA. At the same time, improvements in genome engineering tools have enabled unprecedented control over genetic material. These technologies, combined with a growing understanding of plant systems biology, will irrevocably alter the way we create new crop varieties. As the first wave of genome-edited products emerge, we are just getting a glimpse of the immense opportunities the technology provides. We are also seeing its challenges and limitations. It is clear that genome editing will play an increased role in crop improvement and will help us to achieve food security in the coming decades; however, certain challenges and limitations must be overcome to realize the technology's full potential.

1. INTRODUCTION

1.1 The Role of Genome Engineering in Agriculture

As world population climbs from the current 7.3 billion to a projected 9.7 billion by 2050, there will be an increasing demand to efficiently produce and distribute food. It is predicted that food demand will increase 59%–98% by 2050,[1] which will likely necessitate rethinking current agricultural practices. This challenge—along with higher temperatures, drought, flooding, pests, and diseases—places food security at the top of the international political agenda. Alongside challenges in production, there is an increasing awareness and interest in functional foods—those that have healthier characteristics beyond basic nutrition.[2] Whereas a solution to these challenges is unlikely to come from a single technological advance, it is important to critically evaluate new technologies to determine their role in a solution.

One potential solution to improve food security and enhance food quality relates to the use of genome engineering to create new crop varieties. Genome engineering (or genome editing) can generally be defined as the targeted modification of DNA within living organisms. Due to the wide-ranging utility of modifying an organism's genome, the breadth of applications that fall under the genome engineering umbrella is enormous. Examples of such applications for agricultural purposes can range from basic biology (e.g., understanding gene function) to applied biology (e.g., altering plant structure or characteristics to produce a useful product). In general, the common ground for most genome engineering projects is their reliance on tools that are capable of recognizing and altering a user-selected DNA sequence. This user-selected DNA sequence can include coding regions within genes to noncoding intergenic sequences; the modifications can range from single-nucleotide substitutions to large deletions or insertions. Being able to introduce a wide range of targeted DNA changes, in turn, results in a wide range of potential products, including those that could help address concerns related to food security or quality.

To apply genome engineering to produce useful agricultural products, three major questions need to be addressed: (i) what new, useful traits are to be introduced; (ii) what DNA modifications are required to generate the traits; (iii) how are these modifications physically introduced into a desired crop's genome? Unfortunately, answering these questions can be challenging, particularly for questions (ii) and (iii). For example, limited knowledge of the biology underlying certain complex plant traits (e.g., drought tolerance), and the inability to transform certain crop varieties can create significant bottlenecks when trying to generate new products. Nonetheless, significant progress has been made in applying genome engineering in agriculture. Within the last 5 years, numerous products have emerged from genome-editing platforms, including those with higher yield, drought tolerance, and improved oil characteristics. Here, we review the different types of genome modifications that can be introduced in plants and their potential cellular consequences (Section 2), the successes of applying genome editing in agriculture (Sections 3 and 4), and the current limitations and challenges within this field (Section 5).

2. GENOME EDITING IN PLANTS: POTENTIAL DNA MODIFICATIONS

2.1 Single-Nucleotide Polymorphisms

Perhaps the most subtle targeted genome edit is one that results in a single-nucleotide polymorphism. Here, the total size and organization of the crop genome remains unchanged; however, one nucleotide (out of the millions or billions within a plant's genome) is changed to a different nucleotide. Surprisingly, this subtle change can have profound impacts on cellular function. For example, if positioned appropriately within a gene, a single-nucleotide polymorphism can result in the complete inactivation of gene activity or protein function. This so-called gene knockout can occur when a single-nucleotide polymorphism transforms a codon that normally codes for an amino acid into an early stop codon (Fig. 1B). Having an early stop codon within a gene's coding sequence results in the premature termination of protein synthesis, thereby producing a truncated protein with potentially reduced or no activity.[3–5] Notably, if there are introns downstream of an early stop codon, the mRNA can be subjected to nonsense-mediated decay—a cellular response to aberrantly processed mRNA, which results in degradation of the message.[6] Alternatively, a single-nucleotide polymorphism can destroy protein function if a nonsynonymous substitution is

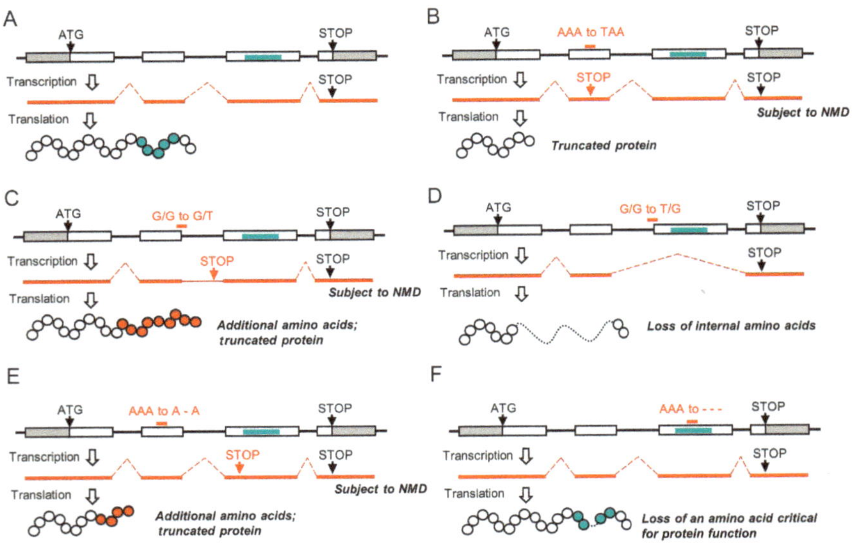

Fig. 1 Examples of genome edits that can alter gene or protein activity. (A) Illustration of transcription and translation of a "normal" gene. (B) Illustration of a single-nucleotide polymorphism leading to an early stop codon. (C) Illustration of a single-nucleotide polymorphism within the conserved 5′ splice site that can result in missplicing and inclusion of an intron. Within this example, the intron is shown to comprise an early stop codon. (D) Illustration of a single-nucleotide polymorphism within the conserved 3′ splice site that can result in missplicing and exclusion of a downstream exon. (E) Illustration of a single-nucleotide deletion that can result in a frameshift and early stop codon. (F) Illustration an inframe deletion within a site that encodes an amino acid critical for protein function. *Gray boxes*, 5′ and 3′ untranslated regions; *white boxes*, exons; *blue rectangle*, DNA region encoding amino acids critical for protein function; *red lines*, mRNA; *white circles*, amino acids; *blue circles*, amino acids critical for protein function; *red circles*, amino acids not normally encoded by the wild-type gene; *NMD*, nonsense-mediated decay; *forward slash*, intron–exon or exon–intron junction; *dash*, deleted nucleotide.

introduced within a codon that normally encodes an amino acid required for protein function.[3,7,8] This substitution, for example, can be within an active site where a substrate molecule binds.

Additionally, a single-nucleotide polymorphism can destroy gene function if it occurs within conserved intron splice/donor sequences. Plant genes contain highly conserved intron splice/donor sequences, which include the 5′ splice site (AG/GTAAG) and 3′ splice site (TGCAG/G).[9–11] Single-nucleotide polymorphisms within these splice sites may significantly affect mRNA processing, and potentially result in production of a nonfunctional protein.[3,12] For instance, mutations in the 5′ splice site can abolish splicing, resulting in inclusion of intron sequence within the mRNA transcript.[13,14] If stop codons are present within the intron, a truncated protein will be produced (Fig. 1C). Alternatively, mutations in the 3′ splice sites can result in exon skipping, wherein a single splicing event removes an exon and flanking intron sequences (Fig. 1D). Also, mutations in splice sites can activate cryptic splice sites, thereby affecting the processing of neighboring exons.[15] An example of where a single-nucleotide polymorphism affected intron splicing and destroyed protein function can be seen with the mutant *FAD3C* gene in canola.[13] FAD3C is a fatty acid desaturase partially responsible for the conversion of linoleic to linolenic acid. A G-to-A base substitution was identified at the 5′ splice site within intron 6 of *FAD3C*. This mutation resulted in impaired splicing, wherein intron 6 was retained in the mature transcript. Stop codons present within intron 6 were predicted to result in early termination of translation and production of a truncated and inactive FAD3C polypeptide. Consistent with this prediction, seed oil from plants with the *FAD3C* mutation had significantly lower linolenic acid content than wild type—a phenotype expected from having a nonfunctional FAD3C protein.

In addition to destroying protein function, single-nucleotide polymorphisms can attenuate or alter protein function. For example, it is possible to dampen or weaken protein activity if a polymorphism results in a missense mutation within a codon that code for an amino acid not entirely critical for protein function.[16,17] Whereas the protein still retains function, the overall activity level has been reduced. On the other hand, a polymorphism can change a protein's properties if a structural change occurs that enables or prevents a certain activity from taking place. For example, a single-nucleotide polymorphism in a gene's coding sequence can result in an amino acid change that prevents molecules from interacting with a certain protein domain. This is seen in the case of the *EPSPS* (5-enolpyruvylshikimate-3-phosphate synthase) gene, which produces a protein involved in branched-chain amino acid synthesis. One or two amino acid substitutions can change the structure of one of the protein's substrate-binding sites, such that the natural substrate continues to bind; however, a chemical analog, glyphosate, is prevented from binding or binding is attenuated.[18]

2.2 Indels

Another class of subtle genome edits are small deletions or insertions (indels), which are typically less than 100 nt. To provide perspective, a deletion of 1–100 nucleotides in a crop plant like maize represents 0.00000004%–0.000004% of the total genome (maize has a 2.5 Gbp genome). However, like substitutions, small indels can have profound impacts on cellular function. For example, gene knockouts can be attained if an indel is introduced into a gene's coding sequence, where the deletion or insertion has a size of $-3N+1$ or $3N-1$ (where N is a whole number), respectively. This type of mutation, also known as a frameshift mutation, frequently leads to early termination of translation and a truncated protein, wherein the truncated protein has attenuated or no activity (Fig. 1E).[19–22] On the other hand, gene knockouts can be attained if the deletion is inframe ($-3N$) and removes codons encoding amino acids critical for protein function (Fig. 1F).[19]

In another application, small indels can be introduced to prevent DNA-binding proteins (e.g., transcription factors) from recognizing their DNA target. One possible use for this approach is to alter expression of a target gene. An example of this application was shown with the *Os11N3* gene in rice. *Os11N3* is a gene that encodes a member of the SWEET sucrose-efflux transporter family. Activation of *Os11N3* (or so-called *OsSWEET14*) gene expression results in export of sucrose from plant cells, which can help support pathogen growth. Gene expression of *Os11N3* is induced by pathogenic transcription factors, which are injected into rice cells by bacteria. One of these pathogen transcription factors binds to a site within the promoter region of *Os11N3*. When small deletions were introduced within the binding site, the transcription factor no longer induced *Os11N3* gene expression, thereby attenuating pathogenicity.[23]

2.3 Large Deletions

Genome editing can also generate large deletions (~100 nt to >1 MB). Introducing large deletions can enable valuable basic plant biology research, as removal of gene clusters, *cis*-regulatory modules, or noncoding RNA genes can provide information on the function of these elements as a group.[24] Alternatively, large deletions can be valuable for applied research, including crop breeding, by providing a means to remove unwanted loci that may be detrimental to fitness or food quality. Notably, introducing large deletions may be one of the most "risky" gene edits, as there is a greater chance to remove genes or functional elements that are necessary for normal plant growth and function.

2.4 Insertions

One of the more practical types of gene edits involves the targeted insertion of DNA sequences into predetermined locations within a plant genome. This is in direct comparison to random insertion using biolistics or *Agrobacterium*-mediated transformation methods, where DNA sequences are randomly integrated into a host's genome. Random integration presents numerous challenges, including (i) the need to screen through numerous events to find those with stable gene expression, (ii) the possibility that the insertion site causes negative effects on agronomic performance, and (iii) the breeding difficulties that arise when there are two or more unlinked transgenes. Targeted insertion overcomes most of these challenges. A DNA sequence of interest, potentially a gene-coding sequence or promoter, is inserted into the host genome, with or without the subsequent removal of host DNA.[25–31] The ability to choose where a gene integrates allows users to select a "safe harbor," where gene expression is reliable and agronomic performance is not affected, or to select a region that already contains desired functional elements for the purpose of trait stacking.

2.5 The Tools

Numerous tools are available to achieve the gene edits described earlier, including sequence-specific nucleases. Sequence-specific nucleases introduce targeted DNA double-strand breaks (DSBs). Normally, DSBs are toxic to cells if unrepaired; therefore, cells have evolved pathways to repair such breaks. In general, plant cells have two major DNA-repair pathways, nonhomologous end joining (NHEJ) and homologous recombination (HR). Both pathways can be exploited to introduce sequence changes into a genome. For example, NHEJ is naturally error-prone, frequently resulting in small indels at the site of repair. Therefore, introducing a DSB at a locus of interest can result in targeted mutagenesis. On the other hand, HR copies information to the break site using a donor DNA template. The donor template is usually a sister chromatid or homologous chromosome. Any sequence differences in the donor will be copied to the broken chromosome. It is possible to "trick" the cell into repairing the break with a user-supplied donor template, thereby allowing user-specified DNA sequence changes to be incorporated into the genome.

Currently, most types of genome edits rely on introduction of DSBs. To introduce a DSB, users have the choice between multiple technologies, including meganucleases, zinc-finger nucleases, TALENs, and CRISPR/Cas systems, with the most popular technologies being CRISPR/Cas and TALENs. Readers are directed elsewhere for reviews about these technologies.[32,33] In addition to nucleases, users also have the option to use Cre-*lox* and other site-specific recombination systems for gene edits,[34] or CRISPR/Cas deaminases, which include Cas9:AID and Cas9:APOBEC1.[35] However, both technologies are limited in the types of genome edits that can be introduced. Deaminases can be used to create targeted SNPs (targeted G→A or C→T changes) at frequencies that are potentially higher than performing nuclease-mediated HR. Cre-*lox* systems are limited to targeted insertion or precise deletion at a location, where necessary recombination sequences are already present. Together, these tools provide users with the ability to introduce diverse types of DNA modifications.

3. AGRICULTURAL DEMANDS AND GENOME EDITING SUCCESSES

3.1 Increasing Food Production

One means to address rising food demands from a growing population is to increase crop yield. However, crop yield is a complex trait governed by many different factors, both environmental and genetic. With regard to the environment, crop yield can be significantly influenced by agricultural practices, which includes nutrient and water supply, and management of weeds and pests. With regard to genetics, crop productivity can be influenced not only by selecting for crops with higher yield but also by selecting for crops that have increased resistance to drought, bacteria, fungi, and viruses, or other factors, such as optimal flowering time. Although the use of genome editing to improve crop productivity is still in its infancy, there have been numerous successes already reported.

3.2 Increasing Crop Productivity

Perhaps the most direct way to increase yield is to increase the productivity of the crop, which includes increasing grain size and weight. Grain weight is mainly attributed to the size of the grain (volume) and the degree of grain filling (plumpness).[36,37] The genetics underlying grain size and weight have been correlated with hundreds of quantitative trait loci (QTLs). Several of the major QTLs have been molecularly characterized, unveiling the underlying functional elements. In rice, these functional elements include G-protein gamma subunit (GS3), serine carboxypeptidase (GS5), protein phosphatase (qGL3/qGL3.1), RING-type E3 ligase (GW2), polyubiquitin-interacting protein (GW5/qSW5), IAA-glucose hydrolase (TGW6), and Snakin/GASA protein (GASR7).[38] In effort to increase yield in rice, CRISPR/Cas9 was used to simultaneously knockout three major genes that negatively regulate grain size: *GW2*, *GW5*, and *TGW6*.[39] The resulting plants displayed an increased grain size and weight between 20% and 30% of wild type. In a different study, CRISPR/Cas was employed in wheat to knockout the three homoeoalleles of *GASR7*.[40] The resulting wheat plants had significantly elevated thousand kernel weight compared to wild-type plants.

Alternatively, yield can be influenced by manipulating flowering time. This was demonstrated in tomato by modifying flower repressor genes. Tomato is classified as a day-neutral plant, although many cultivars flower during long-day conditions.[41,42] Tomato is a sympodial plant with indeterminate plant architecture, wherein the plant continually grows and produces fruit until it dies. Using CRISPR/Cas9, the flowering repressor *SELF-PRUNING 5G* (*SP5G*) gene was knocked out, resulting in a day-length insensitive tomato.[43] The modified plants exhibited early yield due to rapid flowering and compact determinate growth in long-day conditions. In addition to the *SPG5G* gene, mutations were introduced into *SELF PRUNING* (*SP*), known to switch plant architecture from indeterminate to "bushy" determinate.[44] The resulting double mutants displayed rapid flowering bursts and earlier fruit ripening when compared to control lines. Whereas overall yield was lower, harvest index—defined as total yield divided by plant weight—was increased. These results provide information useful to engineer crops for cultivation in different geographical areas, to shorten the cycle of crops to minimize impacts of drought, and to open the possibility of domesticating weed species with agronomical interest.

3.3 Increasing Resistance to Plant Pathogens

Crop losses caused by pathogens, animals, and weeds are estimated to reduce global agricultural productivity between 20% and 40%.[45,46] Efforts to reduce crop losses can indirectly result in an increased yield. Genome editing has been successfully deployed to engineer crops with increased resistance to bacteria and fungi. For example, genome editing was used to improve plant disease resistance in rice to thwart bacterial blight. Bacterial blight, caused by *Xanthomonas oryzae*, is a devastating disease that affects millions of hectares throughout the world. Yield losses in Southeast Asian countries are estimated to range from 10%–20% in moderate conditions to up to 50% in conducive conditions.[47] To facilitate pathogenesis, *X. oryzae* injects effector proteins via a Type III secretion system into the cytoplasm of plant cells. Some of the effector proteins are TAL effectors, which specialize in activating expression of certain genes that help provide a suitable environment for the pathogen to persist. One of these genes, *Os11N3* (also called *OsSWEET14*) is activated when TAL effectors bind to a DNA sequence within the gene's promoter. To disrupt the binding of the *X. oryzae*-derived TAL effectors, TALENs were designed to mutate disease susceptibility elements in the *Os11N3* promoter.[23] Rice plants with mutations in the TAL effector binding elements showed increased resistance to bacterial blight.

Another example of using genome editing to confer disease resistance involved modifying the wheat genome to confer resistance to powdery mildew. Powdery mildew is a fungal disease caused by *Blumeria graminis* f.sp. *tritici* (Bgt).[48] Yield losses range from 10% to 15%, but can reach as high as 40% under heavy epidemics.[49] It was known in barley, Arabidopsis and tomato that the mildew-resistance gene (*MLO*) was associated with increased resistance to Bgt.[50–52] To introduce Bgt resistance in wheat, the three *MLO* homoeoalleles were targeted using TALENs. Wheat plants containing homozygous knockout mutations in all three homoeoalleles displayed heritable, broad-spectrum resistance to powdery mildew.[22]

3.4 Weed Management

Effective weed management is critical to achieve maximum crop growth and productivity. One management method is to spray herbicides that kill or stunt the growth of weeds, but have limited impact on crop plants. Herbicides generally work by binding to and inactivating proteins necessary for normal cellular function. For example, the most widely used herbicide, glyphosate, functions by inactivating the enzyme 5-enolpyruvylshikimate-3-phosphate synthase (EPSPS), thereby shutting down the shikimate pathway. Specifically, glyphosate binds to a pocket within the EPSPS protein that normally binds phosphoenol pyruvate. The binding of glyphosate to the phosphoenol pyruvate site mimics an intermediate state of the ternary enzyme–substrate complex.[53] By preventing the conversion of the substrate shikimate 3-phosphate to 5-enolpyruvylshikimate-3-phosphate, the plant is unable to produce aromatic amino acids required for survival.

Genome editing has been employed to generate plants that are tolerant to herbicides, including glyphosate. One general mechanism for introducing herbicide tolerance involves protein-structure changes, such that the herbicide no longer binds, but the protein remains functional. For example, certain mutations within the phosphoenol pyruvate binding site within EPSPS can prevent glyphosate binding, but still permit phosphoenol pyruvate binding. This was demonstrated in *Linum usitatissimum* (flax) using CRISPR/Cas and single-stranded oligo repair templates to introduce the mutations.[54] Two SNPs were introduced within the *EPSPS* gene, resulting in a T178I and P182A (TIPA) substitution within the phosphoenol pyruvate binding site. When sprayed with glyphosate, edited flax lines showed significantly higher levels of tolerance as compared to wild-type plants. Mutations were also introduced into the rice EPSPS gene using CRISPR/Cas and an intron-replacement strategy.[55] Whereas plants heterozygous for the gene edit had significantly higher levels of glyphosate tolerance, plants containing homozygous gene edits were not viable, suggesting that modification of the endogenous EPSPS gene in rice resulted in insufficient EPSPS protein activity to maintain plant viability.

In addition to EPSPS and glyphosate tolerance, protein-structure changes can be introduced into *acetolactate synthase* (*ALS*) gene(s) for increasing tolerance to ALS-inhibiting herbicides. ALS, also called acetohydroxyacid synthase (AHAS), is a conserved protein required for biosynthesis of the branched-chain amino acids, namely, valine, leucine, and isoleucine. ALS is the primary target for five structurally distinct classes of herbicides, including sulfonylureas (e.g., chlorsulfuron), imidazolinones, triazolopyrimidines, sulfonyl-aminocarbonyl triazolinones, and pyrimidinyl thiobenzoates. Common mutations in ALS that confer resistance to these herbicides occur at positions A122, P197, 205, W574, and S653, in reference to the Arabidopsis ALS protein. There have been several examples of the use of

ZFNs and CRISPR/Cas to introduce mutations in ALS, including in tobacco, soybean, maize, and rice.[56–58] Results from these studies demonstrate the power of genome editing to rapidly introduce herbicide tolerance into plants.

3.5 Increasing Resistance to Abiotic Stress

Other factors influencing yield are abiotic stresses, including drought. Unfortunately, the biological basis of drought tolerance is complex and not well understood, thereby slowing progress toward developing drought-tolerant crops. Nonetheless, one potential method to introduce drought tolerance includes modulating levels of phytohormones, chemicals that control plant growth and development. Of the plant phytohormones, ethylene has been demonstrated to play a role in a wide variety of physiological processes, including stresses due to drought and temperature change.[59] Reducing the sensitivity of the ethylene response in maize resulted in enhanced drought tolerance. Reduced sensitivity was initially achieved in maize by overexpressing *ARGOS* genes—negative regulators of the ethylene response.[60,61] Recently, a genome-editing approach was used to generate drought-tolerant maize.[31] Here, the native maize *GOS2* promoter was inserted upstream of the *ARGOS8*-coding sequence, thereby increasing the expression of *ARGOS8*. Modified plants showed an increase in grain yield by five bushels per acre under flowering stress conditions, with no yield loss under optimal conditions. This work demonstrates not only that genome editing can be used to enhance drought tolerance, but also that a single endogenous gene can be modified to positively affect a complex trait.

4. CREATING HEALTHIER AND MORE NUTRITIOUS FOOD

Whereas increasing food production is paramount, consumers are also looking for improvements in food nutrition and quality. In developing countries, as household incomes rise, consumers demand not only a greater quantity of food but also food of higher quality. Globally, consumers are showing unprecedented interest in how food is produced, and processed, with increased awareness of claims such as "gluten-free," "reduced cholesterol," "excellent source of vitamins and minerals," or "100% whole grain." Later we provide examples of genome editing successes related to food quality.

4.1 Improving Oil Composition

Oils are ubiquitous ingredients in most processed foods and play an important role in food quality and nutrition. Oils high in monounsaturated fats are considered heart healthy and can help reduce bad cholesterol levels. Such oils typically have a longer shelf life and enhanced oxidative stability.[62] In addition, higher intakes of specific dietary fats, including polyunsaturated fatty acids, have been associated with lower risk of heart disease.[63] Unfortunately, high levels of polyunsaturated fats are associated with low oxidative and frying stability. This problem was conventionally solved by partial hydrogenation—a process that reduces levels of polyunsaturated fatty acids but also produces *trans* fats, which have been linked with coronary heart disease and the buildup of plaque in arteries.[64] In June 2015, the Food and Drug Administration announced that partially hydrogenated oils are not "generally recognized as safe" for use in human food. A problem with certain plant oils, including soybean oil, is that they have high levels of polyunsaturated fatty acids and frequently undergo partial hydrogenation to improve stability. Improvements in fatty acid composition within plants may avoid the need for partial hydrogenation, which, in-turn, can result in healthier oils.

In an effort to improve oil composition in soybean, the fatty acid synthesis pathway was modified using TALENs. Soybean oil is primarily composed of palmitic acid (unsaturated), stearic acid (unsaturated), oleic acid (monounsaturated), linoleic acid (polyunsaturated), and linolenic acid (polyunsaturated), wherein these five fatty acids are present in soybean oil at approximately 13%, 4%, 20%, 55%, and 8%, respectively. Two gene families, *FAD2* and *FAD3*, are responsible for the production of most of the polyunsaturated fatty acids found in seed oil: *FAD2* is responsible for converting oleic acid into linoleic acid, and *FAD3* is responsible for converting linoleic into linolenic acid. Using TALENs, two *FAD2* genes, *FAD2-1A* and *FAD2-1B*, were simultaneously knocked out, resulting in soybean oil with significantly higher levels of oleic acid (~80%, compared to ~20% in wild type), and significantly lower levels of both linoleic (~5%, compared to ~55% in wild type) and linolenic acid (~5%, compared to ~8% in wild type).[19] To further improve oil composition, TALENs were used to create knockout mutations in the *FAD3* gene, *FAD3A*, on top of the *FAD2-1A* and *FAD2-1B* mutations.[20] Triple homozygous mutant soybean plants exhibited increased oleic acid levels (82.2%) and reduced linoleic (2.7%) and linolenic acid (2.5%). In addition to soybean, the fatty acid composition of the oil seed plant *Camelina sativa* was modified using genome editing.[65] Here, CRISPR/Cas was employed to knockout the *FAD2* gene, resulting in plants with increased oleic acid (50%, compared to 16% in wild type), decreased linoleic acid (<4%, compared to 16% in wild type), and decreased linolenic (<10%, compared to 35% in wild type). The methods and materials within these studies provide solutions for the demand for oil with increased oxidative stability.

4.2 Reduced-Acrylamide Potatoes

Acrylamide—a suspected human carcinogen—is produced from carbohydrate-rich foods during high-temperature cooking, with a large percentage of intake coming from French fries and potato chips. The accumulation of acrylamide in potato is directly related to storage and processing practices. To ensure a continuous supply of potatoes for direct consumption and processing, potatoes are stored in cold temperatures. During cold storage, potatoes accumulate the reducing sugars glucose and fructose, which are generated from the cleavage of sucrose by the vacuolar invertase protein (VInv).[66] However, subsequent cooking of the potatoes results in a Maillard reaction between the reducing sugars and free amino acids, thereby producing acrylamide. In effort to reduce the accumulation of acrylamide in potatoes, TALENs were used to knockout the *VInv* gene. Nontransgenic potato plants comprising mutations in all four *VInv* alleles were identified. Cold-stored tubers from modified potato plants exhibited lower levels of fructose and glucose, as compared to wild-type plants. Further, after frying, the resulting chips exhibited less browning and a 73% reduction in acrylamide content.[21] Results from this study suggest that genome editing is likely to be an effective approach for minimizing acrylamide formation in processed potatoes.

4.3 Fruits With Increased Antioxidants

One way to improve nutritional characteristics in crops is to increase and stabilize the levels of certain secondary metabolites, including anthocyanins. Anthocyanins are pigments found in leaves, stems, roots, flowers, and fruit tissues, and are known to act as effective antioxidants. Increasing intake of anthocyanins in animal disease models results in increased cardioprotection, life extension in cancer-prone mice, and inhibition of adipocyte development.[67] Anthocyanin production within plant cells is controlled primarily by the activity of transcription factors, particularly the Myb transcription factors. In an effort to increase anthocyanin production in tomatoes, TALENs and CRISPR/Cas were employed to increase the expression of the Myb transcription factor gene, *ANT1*. Here, a strong promoter (CaMV 35S) was inserted upstream of the *ANT1* gene in the tomato variety, Micro Tom.[68] The resulting plants displayed increased levels of anthocyanins as determined by the deep purple pigmentation found in many tissues, including stems, leaves, and fruits.

5. CHALLENGES AND FUTURE OUTLOOK

5.1 Challenges in Delivering Genome-Editing Reagents

One of the most important steps in practicing genome editing in crops, and also one of the most challenging steps, is the delivery of genome-editing reagents and subsequent generation of modified plants. Unfortunately, transformation and regeneration methods for most crops are low efficiency, time consuming and labor intensive, or they are simply unavailable. These shortcomings have created significant bottlenecks in implementing genome editing in crops. Furthermore, each transformation method has limitations and undesirable features, which can reduce the likelihood of obtaining a properly modified plant. The three most popular delivery techniques are *Agrobacterium*, biolistics, and protoplast transformation. Understanding the strengths and limitations of each system is necessary for implementing genome editing in crops.

5.2 *Agrobacterium*

Agrobacterium-mediated transformation has been extensively used for delivering genome-editing reagents. The transformation process begins when plant cells are contacted with *Agrobacterium*. *Agrobacterium* then injects T-DNA/protein complexes into the cytoplasm of the plant cell, and the T-DNA/protein complexes then traffic into the nucleus. Once inside the nucleus, the T-DNA/protein complex can randomly integrate into the host's genome, or it can remain extrachromosomal. If the T-DNA/protein complex remains extrachromosomal, then transient gene expression can be achieved. For most genome engineering applications, T-DNA is stably integrated into the plant genome. Examples of genome editing by stable integration of T-DNA can be found in corn, soybean, barley, *Brassica oleracea*, potato, and rice.[69–77]

Advantages of using *Agrobacterium* to deliver genome-editing reagents include the high probability of achieving single-copy insertions, and protocols have been developed and optimized for transforming many different crops. Disadvantages include: (i) the need for long tissue culture periods to recover transgenic plants, (ii) low frequency of stably transformed plants, (iii) the narrow range of genotypes within a crop species that can be transformed, and (iv) the host-range limitations of certain *Agrobacterium* species.[78] Further, transformation efficiency and regeneration capacity of target cells are usually negatively correlated.[79] Methods to expand the number of genotypes that can be transformed, as well as reducing the need for tissue culture, may help increase the utility of *Agrobacterium* for delivering genome editing reagents to crops.

5.3 Biolistics

Biolistics, or particle bombardment, is a common method for nuclear plant transformation, and for transformations that require DNA to be delivered to chloroplasts and mitochondria. Biolistics is based on the direct delivery of DNA into plant cells using gold or tungsten particles.[80] The gold or tungsten particles are coated with DNA which are shot at plant tissue at high velocity and can become lodged inside plant cells. Once inside the cell, the DNA elutes off the particles and becomes transiently expressed or stably integrates into the host genome. The physical delivery of DNA to cells circumvents host-range limitations, as is sometimes encountered with *Agrobacterium*. Biolistics has been employed to successfully deliver genome-editing reagents in multiple crop plants, including corn, cotton, soybean, and wheat.[22,25,28,58,81]

Advantages of biolistics include: (i) many different tissues and cell types can be transformed, (ii) no binary vector is required, (iii) it is possible to deliver multiple plasmids with high frequencies of cotransformation, (iv) the transformation protocol is simple, (v) large DNA fragments can be delivered, and (vi) the system can be adapted to deliver mRNA or protein.[40,82] Disadvantages include: (i) messy integration patterns, (ii) relatively low throughput, (iii) high input cost, and (iv) the cellular target cannot be controlled (i.e., cytoplasm, nucleus, mitochondria, or plastid).

5.4 Protoplast Transformation

Protoplast transformation involves the direct delivery of DNA to individual plant cells using polyethylene glycol or electroporation. Once inside the nucleus, the DNA can be transiently expressed or stably integrated into the genome. Of the three methods discussed here, protoplast transformation has arguably the highest efficacy, where it is possible to achieve over 70% transformation frequency. Protoplast transformation has been used to successfully deliver genome-editing reagents in multiple crop plants, including potato, wheat, rice, and flax.[21,54,83–85]

Advantages of protoplast transformation include: (i) delivery of multiple plasmids with high levels of cotransformation, (ii) no binary vector required, (iii) high frequency transformation, and (iv) most plant species are amenable to protoplast isolation and transformation. Disadvantages include limited plant species that are amenable to regeneration from protoplasts, and among those that are amenable, the technique is time consuming and labor intensive.

5.5 Overcoming Delivery Limitations

Most transformation methods are limited to one or a few genotypes per crop because only one or a few genotypes are amenable to regeneration. Unfortunately, amenability to transformation and regeneration is usually not correlated with acceptable agronomic performance. In an effort to expand the frequency of transformation and range of genotypes, Lowe *et al.* describe a new method which uses the *Baby boom* (*Bbm*) and *Wuschel2* (*Wus2*) transcription factors.[86] *Agrobacterium*- and biolistic-mediated transformation were employed to deliver *Bbm* and *Wus2* to tissues from maize, rice, sorghum, and sugarcane varieties. Frequencies of transformation increased in rice, sugarcane, and sorghum from 0%–3% to 15%–885%. Whereas this method may not have achieved a complete and universal genotype-independent transformation system for all crops, this is arguably one of the biggest steps in recent years toward this goal.

An alternative means to edit crops that are recalcitrant to transformation is to use virus-based *in planta* delivery systems. Here, a crop plant is infected with an engineered virus that harbors genome-editing reagents. As the virus spreads from cell to cell and throughout the plant, the genome-editing reagents are expressed and introduce genome modifications. Preferably, the virus will spread to meristematic tissue where it can facilitate the introduction of heritable genome modifications. As proof of concept, the geminivirus vector, *Cabbage leaf curl virus*, was used to deliver gRNAs for Cas9 in *Nicotiana benthamiana*.[87] Although high frequencies of mutations were observed in somatic tissues, the use of geminiviruses as delivery vectors precluded production of heritable mutations, as geminiviruses are specifically excluded from meristematic cells. In addition to geminiviruses, an RNA virus, *Tobacco rattle virus* (TRV), was modified to deliver genome-editing reagents to *N. benthamiana* and petunia plants.[88–90] Unlike geminiviruses, TRV has been shown to enter meristematic cells.[91] In agreement with this property, mutations (albeit at low frequencies) were found in seed from plants infected with engineered TRV, demonstrating that TRV is capable of entering meristem cells and delivering genome-editing reagents to create heritable DNA modifications. Improvements in virus-based delivery systems may expand the number of crops that can be edited and improve the frequency that heritable mutations are created.

5.6 Nontransgenic Genome Engineering

When choosing a delivery method, it is also important to consider governmental regulations that may dictate which genome-edited plants can be deployed in the field. In many jurisdictions, introduction of foreign DNA into a cell is a trigger for regulation.[92] Further, the process of delivery may result in undesired off-target modifications due to random insertion of nuclease-encoding DNA. Editing plant genomes by transiently delivering DNA to cells, or by delivering mRNA or protein, may reduce regulatory and off-target concerns related to genome-edited plants. Several approaches have been employed to address this concern, including delivery of sequence-specific nucleases as mRNA,[40,93] protein,[94] mRNA/protein Cas9-gRNA complexes,[57,95] or transient delivery of plasmid DNA or mRNA.[21,40] Results from these studies provide novel approaches to reduce the likelihood of unwanted, random DNA integration.

5.7 Trait Gene/Locus Identification

Another significant challenge for genome editing in agriculture is determining types of genetic modification(s) needed to create a trait of interest. Identifying a target gene or DNA region requires both an understanding of the crop's genome and gene function. However, many of the traits and targets mentioned herein are based on gene-function studies that were performed in model plants. In an effort to provide more information about the genetic makeup of crop plants, many genomes are being, or have been sequenced, including bread wheat, rice, soybean, corn, and potato.[96–100] However, genome sequences are mostly limited to one variety within each crop. Due to genomic differences between varieties, there is a need to resequence target genes or DNA regions, as SNPs, indels or gross chromosomal rearrangements could be present. This resequencing is necessary to carry out accurate genome editing, as using genome editing tools relies on having the precise sequence of the target region in the variety being modified. Further, due to many crops being highly recalcitrant to transformation and regeneration, molecular characterization of gene function has been limited. To help address these concerns, our understanding of gene function and systems biology in crop plants needs to be advanced.

5.8 Frequency of Introducing a Desired Gene Edit

An additional challenge in implementing genome editing is the efficiency that certain edits are introduced. The simplest gene edit, with respect to nuclease technology, is an indel. Indels are created by delivering nucleases to plant cells, wherein the nucleases create targeted DNA DSBs followed by imprecise repair by the NHEJ pathway. Frequency of this type of edit can be extremely high, reaching over 70%.[101] However, indels are only one type of the many edits that can be made Other edits include large deletions, single-nucleotide polymorphisms, gene insertions, gene replacements, and inversions. Frequencies of creating these types of edits are generally much lower than frequencies of creating indels.

Perhaps the most frequent gene edit outside of indels are large deletions, where two DSBs are introduced within the same chromosome, and intervening sequence is deleted. As proof of concept in plants, Zhou *et al.* created two DSBs using CRISPR/Cas in the rice genome to delete large stretches of DNA, approximately 245 kb, at a high frequency (~4/24 transgenic calli).[24] It was later demonstrated in Arabidopsis that the frequency of the deletion is inversely correlated with its size (i.e., the larger the deletion, the lower the efficiency). Whereas plants comprising small deletions of <100 bp were identified at fairly high frequencies (~10%), the frequency of plants with phenotypically selected deletions of 5–120 kb was below 1%.[102]

Low frequencies of gene edits also plague the introduction of single-nucleotide polymorphisms, gene replacements, and gene insertions, as these edits rely on the use of the HR machinery. To achieve HR, both nucleases and donor molecules are provided to plant cells; the resulting DSBs are repaired by the HR machinery using the information within the donor molecule as a template. If the donor molecule has sequence differences from the chromosomal target, the differences will be stably incorporated into the host genome. In general, the efficiency of HR is about two orders of magnitude lower than NHEJ,[103] with reports of frequencies ranging from ~10% to 0.08% of transformed cells.[54,104] Wide ranges of gene targeting frequencies are most likely due to numerous confounding variables, including nuclease efficiency, nuclease type, DSB overhang, plant species, cell type, cell-cycle stage, donor homology-arm length, donor structure (e.g., single stranded vs double stranded vs circular vs linear), composition of nucleotides within the homology arms, the length of sequence being modified, the way the donor was delivered (T-DNA vs plasmid vs virus), and the epigenetic status of the target locus. One solution to address the low frequencies of gene targeting is to establish an efficient transformation pipeline and screen hundreds of transgenic plants—a brute-force approach. However, this is not practical for most research labs, as high-throughput platforms are costly, resource intensive and laborious. An alternative solution is to increase the frequency that gene editing events occur. Qi *et al.* demonstrated in Arabidopsis that knockout of genes involved in classical NHEJ, *ku70*, or *lig4*, resulted in 3- to 16-fold enhancement in HR-based gene targeting frequencies.[105] Further, knockout of a gene involved in HR-based DSB repair using sister chromatids, *smc6b*, resulted in a three- to fourfold enhancement in HR frequencies. It was later shown in rice that concurrent knockout *DNA ligase 4* alongside the delivery of Cas9 and donor molecules, resulted in increased frequencies of HR.[56] Several other strategies to increase HR frequencies have been demonstrated, including overexpression of RecA and RAD54,[106–108] overexpression of the DNA-resecting proteins OsRecQI4 and OsExo1,[109] and delivery of donor molecules on replicons.[68,110–112]

5.9 Breeding Challenges With Genome-Edited Crops

Because only a limited number of genotypes of a given species are amenable to transformation, the majority of gene edits are introduced into varieties with poor agronomic performance. This usually requires genome-edited crops to be placed into breeding programs to transfer the mutations into elite varieties. However, breeding suffers from several drawbacks, including an investment of several years for backcrossing, laborious crossing protocols for certain crops, cost of continual genotyping, and the potential cost and labor of relocating the plants to a different environment (e.g., winter and summer nurseries). Whereas breeding can be a viable option for transferring one or a few mutations, challenges and costs greatly increase as the number of unlinked gene edits increase. For example, when selfing a plant that is heterozygous for a single gene edit, the frequency of progeny comprising homozygous edits is high (1 out of every 4 plants); however, if the plant is heterozygous for four gene edits, the frequency decreases to 1 out of every 256 plants, a frequency that is cumbersome to handle for many breeding platforms. In addition to low frequency challenges, increasing the number of gene edits simultaneously increases the chance that an edit occurs near a locus that negatively effects the overall performance of the plant, thereby creating a linkage drag when crossed with elite varieties.[113] Lastly, for traits such as yield, one needs to carefully compare the value of a product generated by backcrossing to a product that could be generated through a traditional breeding and selection programs, and the annual yield increases associated with such program.

6. CONCLUDING THOUGHTS

Genome editing is poised to change agriculture in the 21st century and beyond. We are already seeing the first wave of genome-edited products emerge, ranging from soybean with improved oil characteristics to drought-tolerant maize. With the continued application of genome editing in crops, we will better understand the constraints and limitations of the technology. These constraints must be addressed before realizing the technology's full potential: two key constraints perhaps being our understanding of plant systems biology and our ability to transform and regenerate elite crop varieties. Reducing or eliminating these constraints gets us closer to the ultimate goal of genome editing in agriculture—having the ability to truly "design" crops from the ground up.

REFERENCES

1. Valin H, Sands RD, van der Mensbrugghe D, et al. The future of food demand: understanding differences in global economic models. *Agric Econ*. 2014;45(1):51–67.
2. Siro I, Kapolna E, Kapolna B, Lugasi A. Functional food. Product development, marketing and consumer acceptance—a review. *Appetite*. 2008;51(3):456–467.
3. Slade AJ, McGuire C, Loeffler D, et al. Development of high amylose wheat through TILLING. *BMC Plant Biol*. 2012;12:69.
4. Gady ALF, Vriezen WH, Van de Wal MHBJ, et al. Induced point mutations in the phytoene synthase 1 gene cause differences in carotenoid content during tomato fruit ripening. *Mol Breed*. 2012;29(3):801–812.
5. Boden SA, Cavanagh C, Cullis BR, et al. Ppd-1 is a key regulator of inflorescence architecture and paired spikelet development in wheat. *Nat Plants*. 2015;1(2):14016.
6. Isshiki M, Yamamoto Y, Satoh H, Shimamoto K. Nonsense-mediated decay of mutant waxy mRNA in rice. *Plant Physiol*. 2001;125:1388–1395.
7. Dierking EC, Bilyeu KD. New sources of soybean seed meal and oil composition traits identified through TILLING. *BMC Plant Biol*. 2009;9:89.

8. Acevedo-Garcia J, Spencer D, Thieron H, et al. Mlo-based powdery mildew resistance in hexaploid bread wheat generated by a non-transgenic TILLING approach. *Plant Biotechnol J.* 2017;15(3):367–378.
9. Brown JWS. Arabidopsis intron mutations and pre-mRNA splicing. *Plant J.* 1996;10(5):771–780.
10. Lorković ZJ, Kirk DAW, Lambermon MHL, Filipowicz W. Pre-mRNA splicing in higher plants. *Trends Plant Sci.* 2000;5(4):160–167.
11. Bilyeu K, Palavalli L, Sleper D, Beuselinck P. Mutations in soybean microsomal omega-3 fatty acid desaturase genes reduce linolenic acid concentration in soybean seeds. *Crop Sci.* 2005;45(5):1830–1836.
12. Sestili F, Palombieri S, Botticella E, Mantovani P, Bovina R, Lafiandra D. TILLING mutants of durum wheat result in a high amylose phenotype and provide information on alternative splicing mechanisms. *Plant Sci.* 2015;233:127–133.
13. Hu X, Sullivan-Gilbert M, Gupta M, Thompson SA. G-to-A mutation at a 5′ splice site of fad3c caused impaired splicing in a low linolenic mutant of canola (Brassica napus L.). *Plant Biotechnol.* 2007;24(4):397–400.
14. Bradley JM, Whitelam GC, Harberd NP. Impaired splicing of phytochrome B pre-mRNA in a novel phyB mutant of Arabidopsis. *Plant Mol Biol.* 1995;27: 1133–1142.
15. McCullough AJ, Baynton CE, Schuler MA. Interactions across exons can influence splice site recognition in plant nuclei. *Plant Cell.* 1996;8(12):2295–2307.
16. Je BI, Gruel J, Lee YK, et al. Signaling from maize organ primordial via FASCIATED EAR3 regulates stem cell proliferation and yield traits. *Nat Genet.* 2016;48(7):785–791.
17. Krieger U, Lippman ZB, Zamir D. The flowering gene SINGLE FLOWER TRUSS drives heterosis for yield in tomato. *Nat Genet.* 2010;42(5):459–463.
18. Sammons RD, Gaines TA. Glyphosate resistance: state of knowledge. *Pest Manag Sci.* 2014;70(9):1367–1377.
19. Haun W, Coffman A, Clasen BM, et al. Improved soybean oil quality by targeted mutagenesis of the fatty acid desaturase 2 gene family. *Plant Biotechnol J.* 2014;12(7):934–940.
20. Demorest ZL, Coffman A, Baltes NJ, et al. Direct stacking of sequence-specific nuclease-induced mutations to produce high oleic and low linolenic soybean oil. *BMC Plant Biol.* 2016;16(1):225.
21. Clasen BM, Stoddard TJ, Luo S, et al. Improving cold storage and processing traits in potato through targeted gene knockout. *Plant Biotechnol J.* 2016;14:169–176.
22. Wang Y, Cheng X, Shan Q, et al. Simultaneous editing of three homoeoalleles in hexaploid bread wheat confers heritable resistance to powdery mildew. *Nat Biotechnol.* 2014;32(9):947–951.
23. Li T, Liu B, Spalding MH, Weeks DP, Yang B. High-efficiency TALEN-based gene editing produces disease-resistant rice. *Nat Biotechnol.* 2012;30(5):390–392.
24. Zhou H, Liu B, Weeks DP, Spalding MH, Yang B. Large chromosomal deletions and heritable small genetic changes induced by CRISPR/Cas9 in rice. *Nucleic Acids Res.* 2014;42(17):10903–10914.
25. Ainley WM, Sastry-Dent L, Welter ME, et al. Trait stacking via targeted genome editing. *Plant Biotechnol J.* 2013;11:1126–1134.
26. Cai CQ, Doyon Y, Ainley WM, et al. Targeted transgene integration in plant cells using designed zinc finger nucleases. *Plant Mol Biol.* 2009;69(6):699–709.
27. Ayar A, Wehrkamp-Richter S, Laffaire JB, et al. Gene targeting in maize by somatic ectopic recombination. *Plant Biotechnol J.* 2012;11:305–314.
28. D'Halluin K, Vanderstraeten C, Van Hulle J, et al. Targeted molecular trait stacking in cotton through targeted double-strand break induction. *Plant Biotechnol J.* 2013;11(8):933–941.

29. Shukla VK, Doyon Y, Miller JC, et al. Precise genome modification in the crop species Zea mays using zinc-finger nucleases. *Nature.* 2009;459(7245):437–441.
30. Zhao Y, Zhang C, Liu W, et al. An alternative strategy for targeted gene replacement in plants using a dual-sgRNA/Cas9 design. *Sci Rep.* 2016;6:23890.
31. Shi J, Gao H, Wang H, et al. ARGOS8 variants generated by CRISPR-Cas9 improve maize grain yield under field drought stress conditions. *Plant Biotechnol J.* 2017;15(2):207–216.
32. Gaj T, Gersbach CA, Barbas CF. ZFN, TALEN, and CRISPR/Cas-based methods for genome engineering. *Trends Biotechnol.* 2013;31(7):397–405.
33. Baltes NJ, Voytas DF. Enabling plant synthetic biology through genome engineering. *Trends Biotechnol.* 2015;33(2):120–131.
34. Wang Y, Yau YY, Perkins-Balding D, Thomson JG. Recombinase technology: applications and possibilities. *Plant Cell Rep.* 2011;30(3):267–285.
35. Li J, Sun Y, Du J, Zhao Y, Xia L. Generation of targeted point mutations in rice by a modified CRISPR/Cas9 system. *Mol Plant.* 2017;10(3):526–529.
36. Xing Y, Zhang Q. Genetic and molecular bases of rice yield. *Annu Rev Plant Biol.* 2010;61:421–442.
37. Sakamoto T, Matsuoka M. Identifying and exploiting grain yield genes in rice. *Curr Opin Plant Biol.* 2008;11(2):209–214.
38. Zuo J, Li J. Molecular genetic dissection of quantitative trait loci regulating rice grain size. *Annu Rev Genet.* 2014;48:99–118.
39. Xu R, Yang Y, Qin R, et al. Rapid improvement of grain weight via highly efficient CRISPR/Cas9-mediated multiplex genome editing in rice. *J Genet Genomics.* 2016;43(8):529–532.
40. Zhang Y, Liang Z, Zong Y, et al. Efficient and transgene-free genome editing in wheat through transient expression of CRISPR/Cas9 DNA or RNA. *Nat Commun.* 2016;7:12617.
41. Binchy A, Morgan JV. Influence of light intensity and photoperiod on inflorescence initiation in tomatoes. *Isr J Agric Res.* 1970;9:261–269.
42. Hurd RG. Long-day effects on growth and flower initiation of tomato plants in low light. *Ann Appl Biol.* 1973;73(2):221–228.
43. Soyk S, Müller NA, Park SJ, et al. Variation in the flowering gene SELF PRUNING 5G promotes day-neutrality and early yield in tomato. *Nat Genet.* 2017;49(1):162–168.
44. Carmel-Goren L, Liu YS, Lifschitz E, Zamir D, Zamir D. The SELF-PRUNING gene family in tomato. *Plant Mol Biol.* 2003;52:1215–1222.
45. Savary S, Ficke A, Aubertot JN, Hollier C. Crop losses due to diseases and their implications for global food production losses and food security. *Food Secur.* 2012;4(4):519–537.
46. Oerke E-C. Crop losses to pests. *J Agric Sci.* 2006;144:31–43.
47. Mew TW, Alvarez AM, Leach JE, Swings J. Focus on bacterial blight of rice. *Plant Dis.* 1993;77:5–12.
48. Glawe DA. The powdery mildews: a review of the world's most familiar (yet poorly known) plant pathogens. *Annu Rev Phytopathol.* 2008;46:27–51.
49. Singh RP, Singh PK, Rutkoski J, et al. Disease impact on wheat yield potential and prospects of genetic control. *Annu Rev Phytopathol.* 2016;54:13.1–13.20.
50. Consonni C, Humphry ME, Hartmann HA, et al. Conserved requirement for a plant host cell protein in powdery mildew pathogenesis. *Nat Genet.* 2006;38(6):716–720.
51. Bai Y, Pavan S, Zheng Z, et al. Naturally occurring broad-spectrum powdery mildew resistance in a central American tomato accession is caused by loss of mlo function. *Mol Plant Microbe Interact.* 2008;21(1):30–39.
52. Piffanelli P, Ramsay L, Waugh R, et al. A barley cultivation-associated polymorphism conveys resistance to powdery mildew. *Nature.* 2004;430(7002):887–891.

53. Schönbrunn E, Eschenburg S, Shuttleworth WA, et al. Interaction of the herbicide glyphosate with its target enzyme 5-enolpyruvylshikimate 3-phosphate synthase in atomic detail. *Proc Natl Acad Sci USA*. 2001;98(4):1376–1380.
54. Sauer NJ, Narváez-Vásquez J, Mozoruk J, et al. Oligonucleotide-mediated genome editing provides precision and function to engineered nucleases and antibiotics in plants. *Plant Physiol*. 2016;170:1917–1928.
55. Li J, Meng X, Zong Y, et al. Gene replacements and insertions in rice by intron targeting using CRISPR-Cas9. *Nat Plants*. 2016;2(10):16139.
56. Endo M, Mikami M, Toki S. Biallelic gene targeting in rice. *Plant Physiol*. 2016;170(2):667–677.
57. Svitashev S, Young JK, Schwartz C, Gao H, Falco SC, Cigan AM. Targeted mutagenesis, precise gene editing, and site-specific gene insertion in maize using Cas9 and guide RNA. *Plant Physiol*. 2015;169(2):931–945.
58. Li Z, Liu Z-B, Xing A, et al. Cas9-guide RNA directed genome editing in soybean. *Plant Physiol*. 2015;169:960–970.
59. Arraes FBM, Beneventi MA, Lisei de Sa ME, et al. Implications of ethylene biosynthesis and signaling in soybean drought stress tolerance. *BMC Plant Biol*. 2015;15(1):213.
60. Shi J, Habben JE, Archibald RL, et al. Overexpression of ARGOS genes modifies plant sensitivity to ethylene, leading to improved drought tolerance in both Arabidopsis and maize. *Plant Physiol*. 2015;169(1):266–282.
61. Guo M, Rupe MA, Wei J, et al. Maize ARGOS1 (ZAR1) transgenic alleles increase hybrid maize yield. *J Exp Bot*. 2014;65(1):249–260.
62. Clemente TE, Cahoon EB. Soybean oil: genetic approaches for modification of functionality and total content. *Plant Physiol*. 2009;151(3):1030–1040.
63. Mozaffarian D, Willett WC. Trans fatty acids and cardiovascular risk: a unique cardiometabolic imprint? *Curr Atheroscler Rep*. 2007;9(6):486–493.
64. Ascherio A, Katan MB, Zock PL, Stampfer MJ, Willett WC. Trans fatty acids and coronary heart disease. *N Engl J Med*. 1999;340:1994–1998.
65. Jiang WZ, Henry IM, Lynagh PG, Comai L, Cahoon EB, Weeks DP. Significant enhancement of fatty acid composition in seeds of the allohexaploid, *Camelina sativa*, using CRISPR/Cas9 gene editing. *Plant Biotechnol J*. 2017;. http://dx.doi.org/10.1111/pbi.12663.
66. Sturm A. Primary structures, functions, and roles in plant development and sucrose partitioning. *Plant Physiol*. 1999;121:1–7.
67. Glover BJ, Martin C. Anthocyanins. *Curr Biol*. 2012;22(5):147–150.
68. Čermák T, Baltes NJ, Čegan R, Zhang Y, Voytas DF. High-frequency, precise modification of the tomato genome. *Genome Biol*. 2015;16:232.
69. Char SN, Unger-Wallace E, Frame B, et al. Heritable site-specific mutagenesis using TALENs in maize. *Plant Biotechnol J*. 2015;13(7):1002–1010.
70. Curtin SJ, Zhang F, Sander JD, et al. Targeted mutagenesis of duplicated genes in soybean with zinc-finger nucleases. *Plant Physiol*. 2011;156(2):466–473.
71. Butler NM, Atkins PA, Voytas DF, Douches DS. Generation and inheritance of targeted mutations in potato (Solanum tuberosum L.) using the CRISPR/Cas system. *PLoS One*. 2015;10(12):1–12.
72. Miao J, Guo D, Zhang J, et al. Targeted mutagenesis in rice using CRISPR-Cas system. *Cell Res*. 2013;23(10):1233–1236.
73. Lawrenson T, Shorinola O, Stacey N, et al. Induction of targeted, heritable mutations in barley and Brassica oleracea using RNA-guided Cas9 nuclease. *Genome Biol*. 2015;16(1):258.
74. Gao H, Smith J, Yang M, et al. Heritable targeted mutagenesis in maize using a designed endonuclease. *Plant J*. 2010;61:176–187.

75. Djukanovic V, Smith J, Lowe K, et al. Male-sterile maize plants produced by targeted mutagenesis of the cytochrome P450-like gene (MS26) using a re-designed I-CreI homing endonuclease. *Plant J.* 2013;76:888–899.
76. Wendt T, Holm PB, Starker CG, et al. TAL effector nucleases induce mutations at a pre-selected location in the genome of primary barley transformants. *Plant Mol Biol.* 2013;83(3):279–285.
77. Gurushidze M, Hensel G, Hiekel S, Schedel S, Valkov V, Kumlehn J. True-breeding targeted gene knock-out in barley using designer TALE-nuclease in haploid cells. *PLoS One.* 2014;9:e92046.
78. Nam J, Matthysse AG, Gelvin SB. Differences in susceptibility of Arabidopsis ecotypes to crown gall disease may result from a deficiency in T-DNA integration. *Plant Cell.* 1997;9(3):317–333.
79. Altpeter F, Springer NM, Bartley LE, et al. Advancing crop transformation in the era of genome editing. *Plant Cell.* 2016;28(7):1510–1520.
80. Sanford JC. The biolistic process. *Trends Biotechnol.* 1988;6(12):299–302.
81. Jacobs TB, LaFayette PR, Schmitz RJ, Parrott WA. Targeted genome modifications in soybean with CRISPR/Cas9. *BMC Biotechnol.* 2015;15:16.
82. Svitashev S, Schwartz C, Lenderts B, Young JK, Mark Cigan A. Genome editing in maize directed by CRISPR-Cas9 ribonucleoprotein complexes. *Nat Commun.* 2016;7:13274.
83. Shan Q, Wang Y, Li J, et al. Targeted genome modification of crop plants using the CRISPR-Cas system. *Nat Biotechnol.* 2013;31(8):686–688.
84. Shan Q, Wang Y, Chen K, et al. Rapid and efficient gene modification in rice and Brachypodium using TALENs. *Mol Plant.* 2013;6(4):1365–1368.
85. Liang Z, Zhang K, Chen K, Gao C. Targeted mutagenesis in Zea mays using TALENs and the CRISPR/Cas system. *J Genet Genomics.* 2014;41:63–68.
86. Lowe K, Wu E, Wang N, et al. Morphogenic regulators baby boom and Wuschel improve monocot transformation. *Plant Cell.* 2016;28:1998–2015.
87. Yin K, Han T, Liu G, et al. A geminivirus-based guide RNA delivery system for CRISPR/Cas9 mediated plant genome editing. *Sci Rep.* 2015;5:14926.
88. Ali Z, Abul-Faraj A, Piatek M, Mahfouz MM. Activity and specificity of TRV-mediated gene editing in plants. *Plant Signal Behav.* 2015;10(10):e1044191.
89. Ali Z, Abul-Faraj A, Li L, et al. Efficient virus-mediated genome editing in plants using the CRISPR/Cas9 system. *Mol Plant.* 2015;8(8):1288–1291.
90. Marton I, Zuker A, Shklarman E, et al. Nontransgenic genome modification in plant cells. *Plant Physiol.* 2010;154(3):1079–1087.
91. Martín-Hernández AM, Baulcombe DC. Tobacco rattle virus 16-kilodalton protein encodes a suppressor of RNA silencing that allows transient viral entry in meristems. *J Virol.* 2008;82(8):4064–4071.
92. Voytas DF, Gao C. Precision genome engineering and agriculture: opportunities and regulatory challenges. *PLoS Biol.* 2014;12(6):e1001877.
93. Stoddard TJ, Clasen BM, Baltes NJ, et al. Targeted mutagenesis in plant cells through transformation of sequence-specific nuclease mRNA. *PLoS One.* 2016;11(5):1–10.
94. Luo S, Li J, Stoddard TJ, et al. Non-transgenic plant genome editing using purified sequence-specific nucleases. *Mol Plant.* 2015;8(9):1425–1427.
95. Woo JW, Kim J, Kwon SI, et al. DNA-free genome editing in plants with preassembled CRISPR-Cas9 ribonucleoproteins. *Nat Biotechnol.* 2015;33(11):1162–1164.
96. Schmutz J, Cannon SB, Schlueter J, et al. Genome sequence of the palaeopolyploid soybean. *Nature.* 2010;463:178–183.
97. Schnable PS, Ware D, Fulton RS, et al. The B73 maize genome: complexity, diversity, and dynamics. *Nature.* 2005;326(5956):1112–1115.

98. Xu X, Pan S, Cheng S, et al. Genome sequence and analysis of the tuber crop potato. *Nature*. 2011;475(7355):189–195.
99. Mayer KFX, Rogers J, Dole el J, et al. A chromosome-based draft sequence of the hexaploid bread wheat (Triticum aestivum) genome. *Science*. 2014;345(6194):1251788.
100. Matsumoto T, Wu JZ, Kanamori H, et al. The map-based sequence of the rice genome. *Nature*. 2005;436:793–800.
101. Li J, Stoddard TJ, Demorest ZL, et al. Multiplexed, targeted gene editing in Nicotiana benthamiana for glyco-engineering and monoclonal antibody production. *Plant Biotechnol J*. 2016;14(2):533–542.
102. Ordon J, Gantner J, Kemna J, et al. Generation of chromosomal deletions in dicotyledonous plants employing a user-friendly genome editing toolkit. *Plant J*. 2017;89(1):155–168.
103. Steinert J, Schiml S, Puchta H. Homology-based double-strand break-induced genome engineering in plants. *Plant Cell Rep*. 2016;35(7):1429–1438.
104. Wright DA, Townsend JA, Winfrey RJ, et al. High-frequency homologous recombination in plants mediated by zinc-finger nucleases. *Plant J*. 2005;44(4):693–705.
105. Qi Y, Zhang Y, Zhang F, et al. Increasing frequencies of site-specific mutagenesis and gene targeting in Arabidopsis by manipulating DNA repair pathways. *Genome Res*. 2013;23(3):547–554.
106. Reiss B, Klemm M, Kosak H, Schell J. RecA protein stimulates homologous recombination in plants. *Proc Natl Acad Sci USA*. 1996;93:3094–3098.
107. Shaked H, Melamed-Bessudo C, Levy AA. High-frequency gene targeting in Arabidopsis plants expressing the yeast RAD54 gene. *Proc Natl Acad Sci USA*. 2005;102(34):12265–12269.
108. Even-Faitelson L, Samach A, Melamed-Bessudo C, Avivi-Ragolsky N, Levy AA. Localized egg-cell expression of effector proteins for targeted modification of the Arabidopsis genome. *Plant J*. 2011;68:929–937.
109. Kwon Y-I, Abe K, Osakabe K, et al. Overexpression of OsRecQl4 and/or OsExo1 enhances DSB-induced homologous recombination in rice. *Plant Cell Physiol*. 2012;53(12):2142–2152.
110. Baltes NJ, Gil-Humanes J, Cermak T, Atkins PA, Voytas DF. DNA replicons for plant genome engineering. *Plant Cell*. 2014;26(1):151–163.
111. Butler NM, Baltes NJ, Voytas DF, Douches DS. Geminivirus-mediated genome editing in potato (Solanum tuberosum L.) using sequence-specific nucleases. *Front Plant Sci*. 2016;7:1045.
112. Gil-Humanes J, Wang Y, Liang Z, et al. High efficiency gene targeting in hexaploid wheat using DNA replicons and CRISPR/Cas9. *Plant J*. 2017;89(6):1251–1262.
113. Peleman JD, van der Voort JR. The challenges in marker assisted breeding. *Cgn*. 2003;125:125–130.

CHAPTER TWO

Use of CRISPR/Cas9 for Crop Improvement in Maize and Soybean

Doane Chilcoat[1], Zhan-Bin Liu, Jeffry Sander
DuPont Pioneer, Johnston, IA, United States
[1]Corresponding author: e-mail address: doane.chilcoat@pioneer.com

Contents

Abstract

CRISPR/Cas enables precise improvement of commercially relevant crop species by transgenic and nontransgenic methodologies. We have used CRISPR/Cas with or without DNA repair template in both corn and soybean for a range of applications including enhancing drought tolerance, improving seed oil composition, and endowing herbicide tolerance. Importantly, by pairing CRISPR/Cas technology with recent advances in plant tissue culture, these changes can be introduced directly into commercially relevant genotypes. This powerful combination of technologies enables advanced breeding techniques for introducing natural genetic variations directly into product relevant lines with improved speed and quality compared with traditional breeding methods. Variation generated through such CRISPR/Cas enabled advanced breeding approaches can be indistinguishable from naturally occurring variation and therefore should be readily accessible for commercialization. The precision, reach, and flexibility afforded by CRISPR/Cas promise an important role for genome editing in future crop improvement efforts.

Progress in Molecular Biology and Translational Science, Volume 149
ISSN 1877-1173
http://dx.doi.org/10.1016/bs.pmbts.2017.04.005

1. INTRODUCTION

Newly developed methods for genome engineering enable plant scientists to make precise changes to the genomes of soybean and corn more efficiently than ever before. These methods use reagents that create double-strand breaks at specific genomic locations. These breaks are then repaired using the cell's own DNA repair machinery. In the absence of an introduced DNA template, the normal error-prone repair process can efficiently generate knockout mutations in targeted genes. When a DNA repair template is provided, prescribed edits can be made, including substitutions, insertions, and deletions of one or more base pairs. Following in the footsteps of the invention of hybrid breeding a century ago, and the introduction of biotech traits a quarter century ago, the ability to precisely edit genomes promises to become an important tool for crop improvement.

First demonstrated at the turn of the 21st century, crop genome engineering has advanced rapidly in the last 5 years, aided by the development of CRISPR/Cas RNA-guided nuclease technology. Before the development of CRISPR/Cas technology, creating targeted double-strand breaks required engineering endonucleases, in order to direct them to cut at intended sites. In contrast, RNA-guided nuclease systems like CRISPR/Cas can be programmed to cut a specific DNA sequence by simply varying ~20 nucleotides of a guide RNA (gRNA) molecule according to simple and well-defined rules of Watson–Crick base pairing. As a result, genome editing has matured from a slow and high-cost process involving considerable reagent development to a simple process that is accessible to both academic and industrial scientists.

Also in the last 5 years, plant tissue culture technology has advanced considerably in important crops such as maize. These new methods enhance the value of genome editing by enabling the direct modification of commercially relevant genotypes. Historically, the inability to efficiently regenerate plants from edited cells in most genotypes limited genome editing to specialized plant lines, selected based on their ability to efficiently regenerate in vitro, not their commercial relevance. As a result, alleles made using genome editing needed to be moved into relevant genetic backgrounds using traditional introgression methods, a lengthy and labor-intensive process. Recently, methods have been developed that enable regeneration of fertile plants from a wide range of commercially relevant maize lines.[1] This is accomplished through overexpression of the cell division promoting transcription factors maize ovule developmental protein 2 (ODP2) and maize Wuschel (WUS) and has potential to enable genotype-independent transformation of most or all monocot crops.

One of the most important applications of genome engineering is the rapid and efficient creation of new varieties that are indistinguishable from those that could be obtained using traditional breeding technologies. One example of this is the creation of null alleles using template-free genome editing. Null alleles are spontaneously generated in nature on an ongoing basis. While such variation has been captured repeatedly throughout the long history crop of improvement (sweet corn is one example), the randomness and slowness of the natural process limits its routine use. Instead of relying on randomly generated mutations, genome editing targets double-strand breaks to a specific gene of interest. When this double-strand break is imprecisely repaired, a loss-of-function mutation can be efficiently generated.

A second example of how genome engineering can be used to generate improved crops that are indistinguishable from those obtained through conventional breeding is the precise transfer of a single allele from one variety to another variety of the same species. For example, genome engineering can be used to move a disease-resistance allele from a disease-resistant nonelite (i.e., noncommercial) variety directly into an elite (i.e., commercial) line. Traditional breeding-based introgression can accomplish this same outcome, but requires an expensive, imprecise, and time-consuming backcrossing process. In contrast, modern genome engineering methods maintain the integrity of the elite genetics by moving only the desired allele quickly and inexpensively. Because these methods generate outcomes that are similar to those obtained through conventional breeding, they are collectively referred to as "advanced breeding technologies." Because advanced breeding generates variation that is indistinguishable from naturally occurring variation, products generated with this process should have a rapid path to commercialization.

In addition to enabling nontransgenic advanced breeding approaches, genome engineering can also be used to develop transgenic products. One such application is the targeted insertion of transgenes. Conventional transformation methods insert transgenes randomly into the genome. While this enables the development of products carrying individual transgenes, the introgression process becomes increasingly time consuming and expensive as the number of transgenic trait loci in a particular product increases. Template-based genome editing enables precise insertion of transgenes into defined locations, enabling traits to be developed within a common genetic "neighborhood." These nearby transgenes can be genetically linked though breeding crosses and then the entire "neighborhood" can be moved as a single linked element into additional elite lines. This enhances the efficiency of trait introgression, removes the cap on the number of traits that can be effectively introgressed, and reduces any plant performance penalties associated with the introgression of flanking sequence regions that accompany the individual genetic loci.

1.1 Experimental Design

In addition to effective double-strand break technology and the ability to regenerate edited cells into fertile plants, successful genome engineering requires careful experimental design. Accurate knowledge of the target genome is needed to select double-strand break sites and to design gRNA sequences. Robust molecular characterization postediting of regenerated plants is necessary to ensure that the intended DNA edits have been made and to determine if any off-site alterations have been made. Importantly, genome engineering requires detailed knowledge of target genes, in order to design edits that improve crop performance.

Comprehensive and accurate reference genomics is the cornerstone of genome engineering experimental design. An accurate and well-annotated reference genome in the variety being edited is needed in order to select target sites, to minimize off-target activity, and to design repair templates. Ironically, the same germplasm diversity that makes crop breeding an effective and fruitful endeavor creates challenges to genome editing, because this diversity means that individual lines vary considerably at the DNA level from one another; in particular, public reference genomes are often quite different from commercial varieties. Furthermore, because the most modern and valuable genetics is always the result of recent breeding efforts, reference genomics is a moving target, requiring ongoing sequencing of new genomes, or at least the generation of new reference sequences for loci being edited. Fortunately, modern DNA-sequencing technologies including those from Illumina and Pacific BioSciences make this possible.

Selection of CRISPR/Cas target sites is a custom process based upon the DNA sequence of the targeted region, the composition of the plant genome, and the experimental goals. The standard approach is to informatically to select CRISPR/Cas target sites that are distinct from other sequences in the genome to reduce the probability that another site is inadvertently coedited.[2] In the case that off-target cutting cannot be avoided, for example, when the same CRISPR–Cas target site occurs in multiple members of a gene family, unintended edits can be easily monitored and crossed out in subsequent generations.

The proximity of the double-strand break to the intended edit is critical for achieving efficient template-based edits. Template-based repair is a local phenomenon and sequence edits are not efficiently incorporated unless a break is targeted nearby.[3] Most successful editing experiments have involved a double-strand break within 100 base pairs of the edited sequence, but additional work is needed to more precisely define the impact of double-strand break proximity to edit location on editing efficiency.

Upon regeneration, individual plants must be molecularly characterized to determine if the intended edits have been made. Preliminary evaluation can be accomplished via sequencing or gel electrophoresis of PCR amplicons. Sequencing of PCR amplicons can effectively characterize genomic deletions and edits introduced from short templates, but is insufficient to confirm edits made with large templates and is inadequate to assess for unintended insertion of any other DNA used in the genome-editing process, such as the CAS9 gene. These uncertainties can be resolved using techniques such as Southern-by-Sequencing (SbS), which uses sequence-based hybridization to capture and deep sequence regions of interest. SbS can both characterize the target locus and confirm the absence of unintended DNA integration.[4]

Genome engineering of corn and soy using CRISPR/Cas is still in its infancy. However, this technology has been used in a variety of applications, including knockout of gene function, introduction of point mutations to modify gene function, removal or addition of regulatory elements to modulate gene expression, and site-specific gene integration at predefined genomic locations for gene stacking.

2. EXAMPLES OF CRISPR/CAS FOR CROP IMPROVEMENT IN MAIZE

The maize genome has been successfully modified using CRISPR/Cas technology.[5–8] Traits that have been altered using this technology include male sterility, lignin biosynthesis, herbicide tolerance, RNA metabolism, secondary metabolism, grain composition, and drought tolerance. The type of editing used to improve these traits includes template-free gene knockout, targeted editing of a small number of codons using a single-stranded or double-stranded DNA as template, insertion of sequences using a large double-strand template bordered by sequence corresponding to the region adjacent to the double-strand break, and replacement of sequences using two cuts that drop out an intervening region, while enabling insertion of a sequence of interest that is provided as a double-strand template bordered by sequence corresponding to genomic regions distal to both cuts. In addition to trait targets, CRISPR/Cas technology has been used to facilitate the stacking of multiple transgenic traits by using homologous recombination to create large numbers of site-specific integration (SSI) landing sites. Template-free gene knockouts are the easiest edits to achieve, followed by targeted transgene insertions containing a selectable marker. DNA template-based experiments using single-stranded DNA oligos or large double-strand DNA molecules can also be successfully achieved with modest resources. Most experiments have used DNA to deliver Cas9, but DNA-free genome editing has been achieved through biolistic bombardment of a preassembled ribonucleoprotein complex.[6]

2.1 Gene Knockout: Waxy Example

The maize WX1 gene encodes a starch synthase involved in kernel composition.[9] Maize homozygous for loss-of-function alleles of the WX1 gene have altered grain starch composition; while wild-type maize grain starch contains about 75% amylopectin and 25% amylose, *Wx1/Wx1* lines have grain starch that is nearly 100% amylopectin.[10] This altered starch composition results in kernels that have a candlewax-like appearance, and therefore is known as "Waxy corn." Cornstarch made from Waxy corn has a number of applications. It is used to improve uniformity, stability, and to provide texture to various food products. Waxy cornstarch also has applications in the textile and adhesive industries and is used as a binder in manufacturing processes in the paper industry. Since the first known reports of Waxy corn in China in 1909,[11] many naturally occurring *Wx1* alleles have been molecularly characterized; several are small deletions affecting the coding sequence.[12]

Waxy hybrids are commercially available and total Waxy acreage in the United States comprises around 0.5 million acres, out of about 90 million total corn acres planted each year. Current Waxy products have two limitations. First, they use older genetics. This is because the *Wx1* alleles are introgressed into elite genetics at the end of the breeding process. Introgression takes years; therefore, Waxy products lag in performance compared to the best conventional products. In addition to this "model year" lag, Waxy hybrids have lower yield than the exact same non-Waxy conventional hybrids. This yield reduction may be the result of imperfect trait introgression, where nonelite alleles near the *Wx1* gene are inadvertently carried along through the introgression process.

Gene deletion using CRISPR/Cas has the potential to address both of these shortcomings. First, genome editing directly in elite germplasm has the potential to be cheaper and faster than trait introgression, allowing the commercialization of modern genetics. Second, *Wx1* deletions generated directly in elite germplasm do not have the imperfections associated with trait introgression, and therefore may not have yield drag.

CRISPR/Cas was used to delete the entire WX1 gene using a template-free two Cas9–gRNA approach. One Cas9–gRNA targeted a sequence upstream of the transcriptional start site and another Cas9–gRNA-targeted cutting downstream of the stop codon (Fig. 1). The region between the two targets was excised and the remaining distal chromosomal ends were joined through the nonhomologous end joining (NHEJ) pathway for DNA damage repair resulting in a *Wx1* null allele with the Waxy phenotype. Genome modification was achieved using particle bombardment of immature embryos with gold particles carrying six different plasmids encoding two gRNAs, the *Streptococcus pyogenes* CAS9 gene, the maize ODP2 gene, and the maize WUS gene, and the NPT2 selectible marker.[1] The NPT2 gene encodes a phosphotransferase protein that provides resistance to kanamycin and G418.[13]

Regeneration frequency was 2%–29% across genotypes (i.e., 2–29 plants were regenerated from every 100 embryos bombarded). Regenerated plants were analyzed for the presence of the desired edits using PCR amplification of the targeted region followed by Illumina-based DNA sequencing. On average 10% of regenerated plants contained deletions between the two

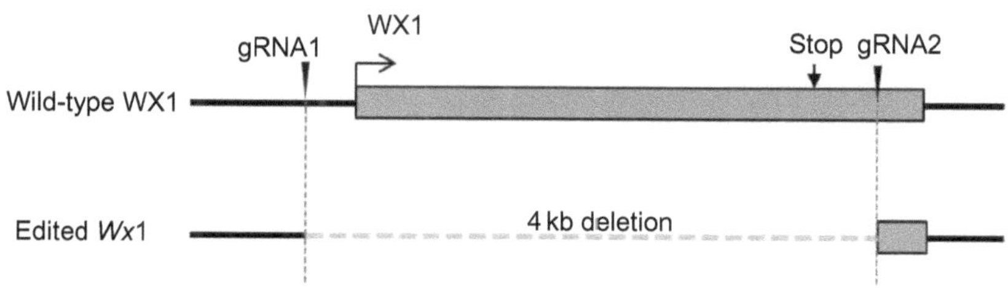

Fig. 1 Design of *Wx1* gene knockout. The WX1 transcript is represented by a *thick gray bar* with the transcript start site is denoted by a *right facing arrow*. Cut sites for Cas9–gRNA targets are represented as *black arrows*. The ~4 kb region that is deleted in the desired *Wx1* edit is denoted by a *dotted gray line*.

Cas9–gRNA targets, while 80% contained only mutations at individual Cas9–gRNA target sites. About 10% of plants containing a deletion were biallelic, with deletions in both WX1 alleles. Plants carrying the desired deletions between the two Cas9–gRNA sites were outcrossed to the inbred used in transformation and the progeny were self-pollinated or outcrossed. In the following generation, plants were selected to contain the desired edit and none of the other bombarded plasmid DNA. The absence of plasmid DNA was initially determined using quantitative PCR designed against elements in each of the six plasmids used in bombardment. This was followed by SBS,[4] which is capable of detecting even very small fragments of the introduced DNA. Of the plants carrying deletions, more than 80% of the *Wx1* mutations could be segregated away from the selectable marker and all other plasmid DNA. As expected, self-pollination of hemizygous plants produced seeds that segregated 1:3 for the recessive Waxy phenotype. In total, *Wx1* deletions were generated in more than 10 different elite inbreds, which enables direct production of CRISPR/Cas-derived Waxy hybrids. The net efficiency of this process, including transformation frequency, molecular attrition, and greenhouse attrition, was about 0.5%; i.e., one line containing the desired mutation and no helper DNA was established for every 200 immature corn embryos bombarded. However, efficiency varied considerably among experiments.

Preliminary characterization of these CRISPR/Cas-derived lines identified no change to phenotype other than the characteristic Waxy trait. Starch composition from these lines was quantitatively determined using established methods and found to be indistinguishable from previously characterized *Wx1* null alleles.

2.2 Gene Edit: Acetolactate Synthase Example

Sulfonylurea herbicides prevent branched amino acid biosynthesis in plants by inhibiting acetolactate synthase (ALS).[14–16] Resistance to the sulfonylurea herbicide chlorsulfuron can be achieved by mutating a single amino acid in the ALS gene from proline to serine.[14,17] Maize contains two ALS genes, ALS1 and ALS2.[18] Although these genes are very similar, nucleotide differences allow the design of Cas9–gRNA targets specific to one or the other. ALS2 was specifically targeted using a Cas9–gRNA that directs cutting 26 nucleotides upstream of the target edit that changes a codon to one for serine instead of proline (Fig. 2).[5]

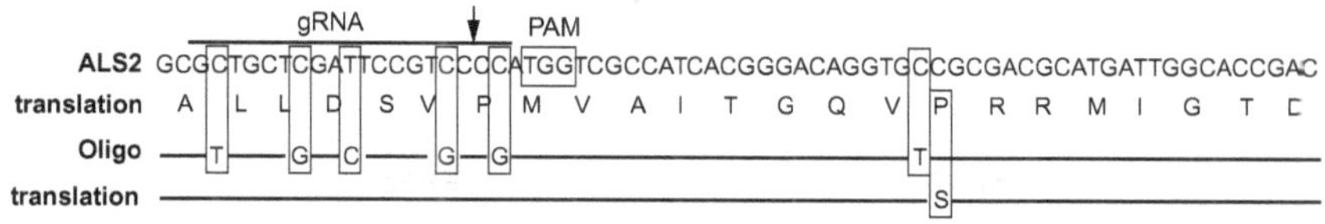

Fig. 2 Design of ALS2 gene edit. A Cas9–gRNA targets cuts 26-bp upstream from the intended edit site. The oligo repair template introduces the nonsynonymous edits changing the CCG–Proline codon to a TCG–Serine codon as well as several silent mutations (*boxed*) in the Cas9–gRNA target site to prevent recutting after template repair.

ALS gene editing was achieved using particle bombardment of immature embryos with gold particles carrying a 127-nucleotide single-strand DNA oligo repair template and five plasmids carrying: a gene encoding the gRNA, the *S. pyogenes* CAS9 gene, a MoPAT-DsRED selectable marker gene, the maize ODP2 gene, and the maize WUS gene. The MoPAT-DsRED marker creates resistance to the herbicide bialaphos and also creates cells with red color.[19] In these experiments, bialaphos selection was not used but the red color was used to confirm successful DNA delivery.

Approximately 800 immature embryos were bombarded and transferred to media containing chlorsulfuron in order to directly select for the desired edits. Three independent plants were regenerated from resistant calli; all contained the intended *Als2* edits. ALS1 remained unchanged, as determined by sequencing of PCR amplicons spanning the site of editing. Hemizygous plants carrying the intended mutation were self-pollinated. Segregating seeds were planted in flats and sprayed with chlorsulfuron after 4 weeks. As expected, plants carrying the edited *Als2* allele exhibited resistance to chlorsulfuron.

In this experiment, a short (127 bp) single-strand DNA oligo was used as the editing repair template. Similar results were obtained using a larger (794 bp) double-strand DNA fragment. Chlorsulfuron resistant plants were also generated by cobombarding immature embryos with the 127 bp single-strand oligo as the only introduced DNA and a preassembled Cas9–gRNA complex.[6]

2.3 Promoter Swap: ARGOS8 Example

Drought tolerance is an important attribute of commercial maize.[20] The plant hormone ethylene is known to play a key role in drought tolerance; in particular, lower ethylene levels result in higher drought tolerance. The maize ARGOS8 gene negatively regulates the ethylene response by encoding a protein that interacts with the ethylene receptor signaling complex.[21] Wild-type ARGOS8 gene expression is low and largely absent from nonreproductive tissues. Moderate constitutive transgenic over-expression of ARGOS8 has been shown to negatively regulate the ethylene response and to improve drought tolerance.[7]

CRISPR/Cas gene editing was used to alter expression of the endogenous ARGOS8 gene by replacing the native promoter with the maize GOS2 promoter, which normally drives moderate constitutive expression of the SUI1 transcription factor.[22]

This promoter swap was achieved using particle bombardment of immature embryos with gold particles carrying a DNA repair template consisting of the GOS2 promoter flanked by two DNA fragments of approximately 400 bp matching the genomic sequences immediately adjacent to the Cas9 cleavage sites in the ARGOS8 locus (Fig. 3) and five plasmids carrying: each of the two gRNAs, a gene encoding the *S. pyogenes* Cas9, two plasmids carrying maize ODP2 and maize WUS, and a gene encoding the phosphomannose isomerase (PMI) selectable marker.[21] The PMI selection system utilizes the PMI gene encoding for PMI that converts mannose-6-phosphate that is toxic to maize cells to nontoxic fructose-6-phosphate.[23]

Of approximately 3000 immature embryos particle bombarded, 334 shoots were placed on regeneration media containing mannose-6-phosphate, of which 172 developed into plantlets. 23 of the 172 plantlets obtained from the initial screening were positive for at least one junction. Among them, three were positive for both junctions, two of which developed into fertile plants. One regenerated plant was free of any DNA element used in transformation and was further characterized and evaluated through field testing.

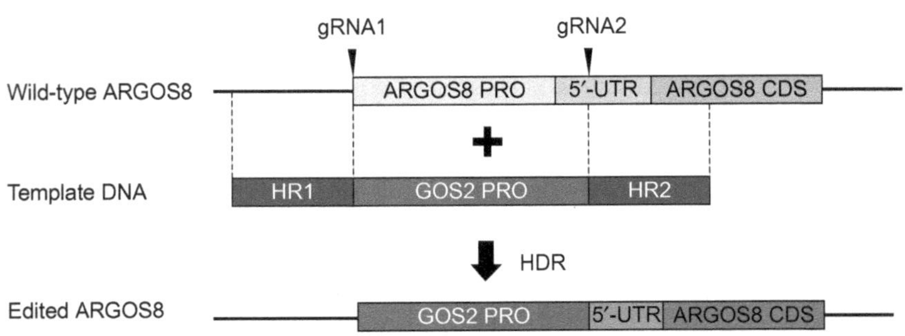

Fig. 3 Design of ARGOS8 promoter swap. One Cas9–gRNA is designed to target cutting upstream of the putative AGROS8 promoter region and a second targets cutting in the ARGOS8 5′ untranslated region. The donor template consists of the GOS2 promoter flanked by ~400 bp sequences corresponding to the regions flanking the target breaks.

The genome-edited ARGOS8 variant was field tested as a hybrid by crossing to an inbred tester. This hybrid was evaluated against a comparable wild-type hybrid under both well-watered and drought stress environments. As expected from previous agronomic characterization of transgenic GOS2-ARGOS8, the edited hybrids performed similarly to wild type under well-watered conditions, but had increased yield when grown under drought stress.[7]

2.4 Gene Insertion: Complex Trait Locus Example

The vast majority of corn grown in the Americas contains multiple transgenic traits, typically including resistance to two or more herbicides and multiple *Bt* genes that provide resistance to various insect pests. It is not uncommon for a commercial product to contain five different transgenic traits. These traits were generated using random transformation and are located throughout the genome. When a new inbred is developed through breeding, transgenic traits must be added to it through the process of trait introgression. This introgression process includes crossing a transgenic donor line with a new elite inbred followed by repeatedly backcrossing the resulting progeny with the elite line in order to recover the elite line's genome plus a small insertion containing the transgene.[24] Trait introgression is effective, but has three shortcomings[1]: it is costly and time consuming, requiring multiple generations and several years[2]; some unintended donor DNA is always inadvertently present in the final product, often including several million base pairs adjacent to the transgene[3]; it is costly and slow to assemble a large number of traits, making it impractical to introgress more than around six unlinked loci.

The obvious solution to this problem is to colocate transgenes so that they behave as a single genetic locus. Because the limitations of trait introgression are primarily related to the number of loci being introgressed, colocation of transgenes has the potential to enable commercial products with any number of transgenic traits. Colocation can be accomplished by creating DNA constructs that contain multiple transgenes, so-called molecular stacks. This strategy has been widely employed for many years. For example, the TC1507 maize event trait contains both a *Bt* gene that provides insect tolerance and an acetyltransferase gene that confers herbicide tolerance. While useful, this approach has several limitations. First, transformation efficiencies are low when large constructs containing multiple genes are used. This limits the number of genes the can be combined. Second, nearby genes (e.g., within a few kb) have the potential to impact one another's expression.

These potential gene–gene interactions within an event complicate the design and discovery of optimized molecular stacks.[25] Third, individual components of molecular stacks cannot be swapped in and out by genetic crossing—if a particular functionality is no longer needed it cannot be individually removed, and a new functionality cannot be added to the molecular stack. An alternative to molecular stacks is the development of complex trait loci (CTLs). CTLs are similar to molecular stacks in that several transgenes are genetically colocated. But instead of only a few hundred or thousand base pairs of separation, the transgenes are separated by 50 kb to ≥1 Mb, while still being closely linked genetically (within a few centimorgans). Being greater than 50 kb away[1] mitigates potential interaction between transgenes and[2] allows old traits to be removed and new traits to be added through genetic recombination, while still allowing the transgenes to be treated as a single locus during trait introgression.

In order to create a maize CTL, CRISPR/Cas genome editing can be used to target transgenes into selected genomic regions through homology-directed repair. However, at present, this is a relatively inefficient process. In contrast, SSI using the FLP/FRT recombinase system is an extremely efficient process.[26] One reason SSI is so efficient is that it utilizes a chromosomal promoter trap coupled with a DNA construct containing a promoter-less selectable marker (Fig. 4). In this system, random integration of the DNA construct does not confer resistance to the selectable marker, and so most of the events that are recovered are genuine SSI products, not the result of random integration. The central limitation of SSI is that it requires a preintegrated target site containing a promoter trap and FRT recombination sites, whereas CRISPR/Cas allows direct targeting anywhere in the genome.

We have combined the best attributes of these two technologies, using CRISPR/Cas to insert a SSI target construct at selected locations in the maize genome and SSI to test various transgenic traits at these locations. Target locations were selected to be near one another, but far enough apart (≥0.1 centimorgans) to be genetically separable, and were selected to be distant from endogenous genes. Several dozen SSI target lines were generated using biolistic bombardment with plasmids encoding a gRNA, a gene encoding the *S. pyogenes* CAS9 gene, maize ODP2, maize WUS, and a double-strand DNA template containing about 400 base pairs of sequence corresponding the genomic regions on in either side of the cut site with a NPT2 selectable marker bordered by FRT sites in the middle. Events were regenerated on media containing G418. Several dozen lines were generated that had exactly the intended DNA sequence, as evaluated by SbS, with an

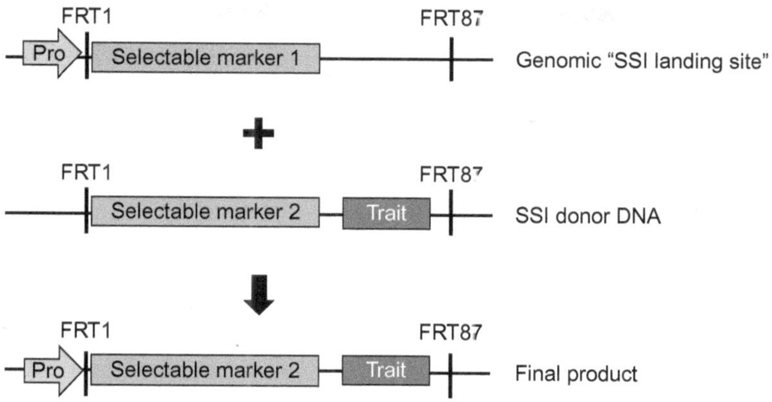

Fig. 4 Design of a SSI landing site. Genomic SSI landing sites containing a FRT promoter trap are introduced at target CTL locations using CRISPR/Cas. Lines containing these landing sites can be retransformed with the FLP recombinase and a donor DNA containing a new selectable marker and the desired traits. This results in the final product, containing the new selectable marker and trait gene(s).

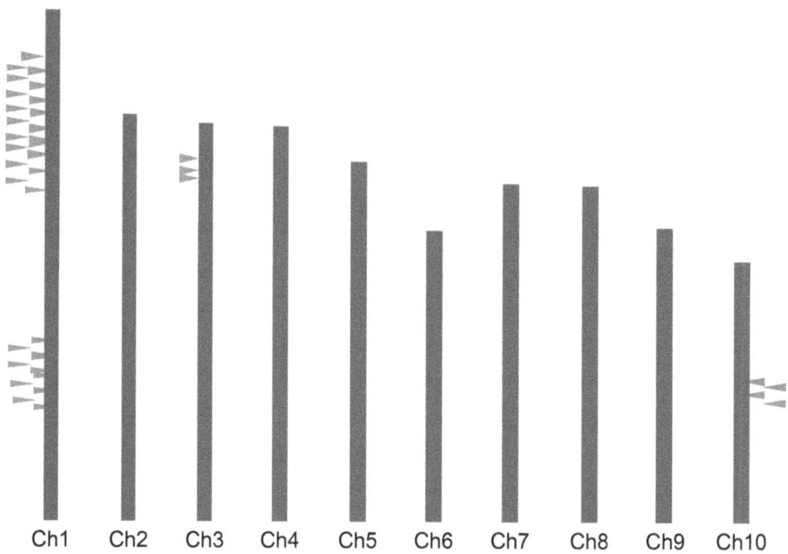

Fig. 5 Complex trait loci landing sites *in Zea mays* genome. Several dozen FRT landing sites were inserted in four genomic "neighborhoods" throughout the *Z. mays* genome. Each neighborhood spans 1–5 centimorgans.

average efficiency of about 0.1% (i.e., 1 perfect event per 1000 embryos bombarded). A subset of these lines were characterized for genetic location, transgene expression level, transformability, and agronomics. These SSI target lines can now be used to generate CTLs (Fig. 5).

3. EXAMPLES OF CRISPR/CAS FOR CROP IMPROVEMENT IN SOYBEAN

CRISPR/Cas has also been used to introduce targeted edits within the soy genome.[27] Traits that have been altered using this technology include seed oil and protein composition, and herbicide tolerance. Similar to maize, the types of editing used to improve these traits include template-free gene knockout, targeted editing of a small number of amino acids using a double strand DNA fragment as template, and insertion of entire genes using a large double-strand template bordered by regions of homology.

3.1 Gene Knockout: Multiplexing Example

CRISPR/Cas can be efficiently used to knockout genes in soy through biolistic bombardment of embryogenic tissue with DNA encoding the *S. pyogenes* Cas9 and one or more genes encoding gRNA(s), followed by plant regeneration with hygromycin as the selectable marker.[26] Gene knockout using CRISPR/Cas is so efficient at creating null alleles that multiple genes can be simultaneously edited in a single experiment. In one approach, multiple genes that share a common sequence such as paralogous family members can be targeted with a single gRNA. Alternatively, multiplexing can be achieved by using two or more gRNAs each targeted to a different DNA sequence. Using a combination of these approaches, five gRNAs targeted to nine genes were codelivered by particle bombardment into embryogenic cultures from which 200 plants were regenerated. Molecular analysis revealed edits in all nine genes with edit efficiencies ranging from 8% to 76% (Table 1). Both monoallelic and biallelic edits were recovered and three plants had edits in all nine genes (Table 2).

3.2 Gene Edit: ALS

Similar to maize, soy ALS genes can be edited to confer resistance to sulfonylurea herbicides.[28] Soy contains four ALS genes, ALS1, 2, 3, and 4. ALS1 on chromosome 4 was specifically targeted using a Cas9–gRNA based on the unique PAM sequence or the unique nucleotide sequence at the #1 position upstream of the PAM in different soybean ALS genes (Fig. 6). The proline at position 178 was edited to serine by changing the codon from CCC to AGC on one allele of the ALS1 gene on chromosome 4 using a ~1 kb double-strand DNA repair template. The edited P178S ALS1 allele

Table 1 Results for Multiplex Gene Knockout Experiment

	Gene1	Gene2	Gene3	Gene4	Gene5	Gene6	Gene7	Gene8	Gene9
Total plants screened	200	200	199	200	200	200	200	200	200
WT	75	116	174	178	83	119	48	182	184
Monoallelic edit	48	41	23	18	44	64	28	17	15
Biallelic edit	77	43	2	4	73	17	124	1	1
Monoallelic edit (%)	24.0	20.5	11.6	9.0	22.0	32.0	14.0	8.5	7.5
Biallelic edit (%)	38.5	21.5	1.0	2.0	36.5	8.5	62.0	0.5	0.5
Total edit (%)	62.5	42.0	12.6	11.0	58.5	40.5	76.0	9.0	8.0

"WT" is the number of plants that contain two wild-type alleles for that gene of interest. The "monoallelic edit" is the number of plants with only a single-edited allele. The "biallelic edit" is the number of plants that have both alleles edited.

Table 2 Editing Profile of Nine Plants From Four Independent Regenerants

Plant ID	Gene1	Gene2	Gene3	Gene4	Gene5	Gene6	Gene7	Gene8	Gene9
1	Biallelic	Biallelic	Monoallelic	Biallelic	Biallelic	Monoallelic	Biallelic	Biallelic	Biallelic
2	Biallelic	Biallelic	Monoallelic	Biallelic	Biallelic	Monoallelic	Biallelic	Biallelic	Monoallelic
3	Biallelic	Biallelic	Monoallelic	WT	Biallelic	Biallelic	Biallelic	Monoallelic	Monoallelic
4	Biallelic	Biallelic	Monoallelic	WT	Biallelic	Biallelic	Biallelic	Monoallelic	Monoallelic
5	Biallelic	Biallelic	Monoallelic	WT	Biallelic	Biallelic	Biallelic	Monoallelic	Monoallelic
6	Biallelic	Biallelic	WT	Monoallelic	Biallelic	Biallelic	Biallelic	Monoallelic	Monoallelic
7	Biallelic	Biallelic	WT	Monoallelic	Biallelic	Biallelic	Biallelic	Monoallelic	Monoallelic
8	Biallelic	Biallelic	Monoallelic	Monoallelic	Biallelic	Biallelic	Biallelic	Monoallelic	Monoallelic
9	Biallelic	Biallelic	WT	Monoallelic	Biallelic	Biallelic	Biallelic	Monoallelic	Monoallelic

Plants 1 and 2 are sibling plants from the same regenerant. Plants 3, 4, and 5 are sibling plants from a second regenerant. Plants 6, 7, and 8 are sibling plants from a third regenerant. Plant 9 is from a fourth regenerant.

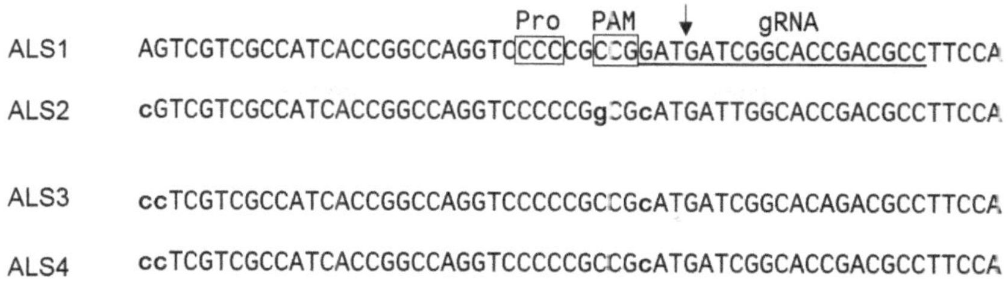

Fig. 6 Sequence alignment of four wild-type ALS genes around the proline at position 178 (*box*). The ALS1 gRNA target (complementary strand) is *underlined* and the PAM (complementary strand) is identified in the *adjacent box*. The expected cleavage site is indicated with an *arrow*. The sequence difference for the other soybean ALS genes as compared to ALS1 is indicated by *bold lowercase letters*.

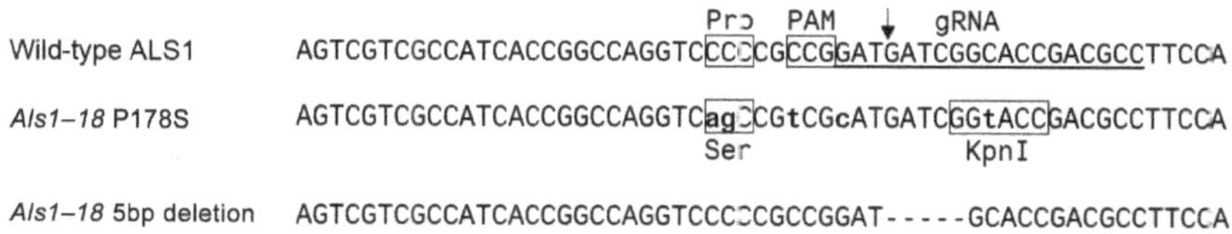

Fig. 7 Generation of edits to soybean ALS1 gene. The edited P178S *Als1–18* gene contained five designed mutations (*bold, lowercase*) introduced by Cas9–gRNA-directed template repair. These mutations changed one amino acid from P178 to 178S. Additional silent modifications were introduced to create a KpnI recognition site (*box*) and to alter the target site to prevent recutting after template repair. The second allele contained a 5-bp deletion (*dashes*).

provided resistance to chlorsulfuron in the regeneration process.[27] In the same event, the second allele of the ALS1 gene was present and contained a 5-bp deletion near the Cas9–gRNA cleavage site (Fig. 7).

3.3 Gene Insertion: Complex Trait Locus

As with corn, multiple genomic regions have been selected for CTL development in soybean using CRISPR/Cas. In total, 14 unique gRNAs were designed and tested for seven target regions. Expression cassettes containing gRNAs, the *S. pyogenes* Cas9 gene, and a donor DNA (repair DNA) cassette were codelivered in the bombardment-based transformation process.[27] The donor DNA construct contained a soybean *S*-adenosylmethionine synthetase gene promoter expressing the hygromycin phosphotransferase gene HPT to confer hygromycin resistance flanked by ∼1 kb fragments derived from the sequences flanking the genomic Cas9–gRNA target site. Yeast FLP recombinase recognition sites (FRT) sites were placed upstream and downstream of the coding sequence in the hygromycin resistance gene to create a promoter trap to enable future gene targeting by SSI.[26,29]

Efficient site-specific cleavage was detected for all 14 gRNAs and SSI landing site integrations were detected in all seven target sites. The gene integration efficiencies ranged from 1% to 2% (based on the number of stable events screened) for the different Cas9–gRNA target sites. As with corn, these results indicate that CRISPR/Cas technology will be widely applicable in soy.

4. LOOKING FORWARD

The last few years have seen tremendous technical progress in genome engineering. Watershed advances in reference genomics, efficient and robust germplasm-independent plant transformation and regeneration technology, robust processes for molecular characterization of genome edits, and the development of CRISPR/Cas technology have now come together to enable genome engineering in maize and soy.

Limitations of the current technology include target site flexibility and overall process efficiency, although neither of these limitations has, as far as we are aware, prevented any product development. Target site flexibility may be improved through the development of RNA-guided nucleases with different properties such as different PAM sequences or different gRNA length requirements.[30] Process efficiency may be improved by the transient delivery of preassembled Cas9–gRNA ribonucleoproteins[6] or Cas9 mRNA/gRNA,[31] by protocols that increase the frequency of homology-directed repair, or by improvements in regeneration efficiency.

Now that genome editing can be used to modify the genomes of corn and soy, the key open questions are around social acceptance, governmental regulation, and trait targets. How will society regard crop varieties that are generated using advanced breeding technologies, varieties that are completely indistinguishable from those generated using conventional breeding technologies, but whose development made use of modern biotechnology? How will global regulatory agencies deal with these new varieties? Last, what genomic targets should be engineered to improve crop performance, and what gaps in our understanding of fundamental biology need to be addressed in order to design the most useful genome modifications? Resolving these questions is critical if we are to realize the full potential of CRISPR/Cas genome engineering in corn and soy.

ACKNOWLEDGMENTS

We thank Melanie Dana for help preparing this chapter. The work described here was supported by the Pioneer Applied Technology Systems organization including Vector Construction, Crop Genome Engineering, Controlled Environments, and Nucleic Acid Analysis. Individual projects were led by Josh Young, Sergei Svitashev, Scott Betts, Bob Meeley, Huirong Gao, Jinrui Shi, Mark Cigan, Dave Peterson, Keith Lowe, Bill Gordon-Kamm, Zhongsen Li, Michael Lassner, Carl Falco, Chris Scelonge, and Emily Wu.

REFERENCES

1. Lowe K, Wu E, Wang N, et al. Morphogenic regulators baby boom and Wuschel improve monocot transformation. *Plant Cell*. 2016;28(9):1998–2015.
2. Graham DB, Root DE. Resources for the design of CRISPR gene editing experiments. *Genome Biol*. 2015;16:260.
3. Eliott BRC, Winderbaum J, Nickoloff JA, Jasin M. Gene conversion tracts from double-strand break repair in mammalian cells. *Mol Cell Biol*. 1998;18:93–101.
4. Zastrow-Hayes GM, Lin H, Sigmund AL, et al. Southern-by-sequencing: a robust screening approach for molecular characterization of genetically modified crops. *Plant Genome*. 2015;8. http://dx.doi.org/10.3835/plantgenome2014.08.0037.
5. Svitashev S, Young JK, Schwartz C, Gao H, Falco SC, Cigan AM. Targeted mutagenesis, precise gene editing, and site-specific gene insertion in maize using Cas9 and guide RNA. *Plant Physiol*. 2015;169(2):931–945.
6. Svitashev S, Schwartz C, Lenderts B, Young JK, Mark CA. Genome editing in maize directed by CRISPR-Cas9 ribonucleoprotein complexes. *Nat Commun*. 2016;7:13274.
7. Shi J, Gao H, Wang H, et al. ARGOS8 variants generated by CRISPR-Cas9 improve maize grain yield under field drought stress conditions. *Plant Biotechnol J*. 2017;15(2): 207–216.
8. Char SN, Neelakandan AK, Nahampun H, et al. An agrobacterium-delivered CRISPR/Cas9 system for high-frequency targeted mutagenesis in maize. *Plant Biotechnol J*. 2017;15(2):257–268.
9. Nelson OE, Rines HW. The enzymatic deficiency in the waxy mutant of maize. *Biochem Biophys Res Commun*. 1962;9(4):297–300.
10. Sprague GBB, Hixon R. Some effects of the waxy gene in corn on properties of the endosperm starch1. *Agron J*. 1943;35(9):817.
11. Collins G. A New Type of Indian Corn From China. Washington: Government Printing Office; 1909:1–30. *(Bulletin) Bureau of Plant Industry*; vol 161. http://library.umac.mo/ebooks/b28370405.pdf.
12. Wessler S, Tarpley A, Purugganan M, Spell M, Okagaki R. Filler DNA is associated with spontaneous deletions in maize. *Proc Natl Acad Sci USA*. 1990;87(22):8731–8735.
13. Fraley RT, Rogers SG, Horsch RB, et al. Expression of bacterial genes in plant cells. *Proc Natl Acad Sci USA*. 1983;80(15):4803–4807.
14. Haughn GW, Smith J, Mazur B, Somerville C. Transformation with a mutant Arabidopsis acetolactate synthase gene renders tobacco resistant to sulfonylurea herbicides. *Mol Gen Genet*. 1988;211(2):266–271.
15. Lee KY, Townsend J, Tepperman J, et al. The molecular basis of sulfonylurea herbicide resistance in tobacco. *EMBO J*. 1988;7(5):1241–1248.
16. Yu Q, Powles SB. Resistance to AHAS inhibitor herbicides: current understanding. *Pest Manag Sci*. 2014;70(9):1340–1350.
17. Walter KL, Strachan SD, Ferry NM, Albert HH, Castle LA, Sebastian SA. Molecular and phenotypic characterization of Als1 and Als2 mutations conferring tolerance to acetolactate synthase herbicides in soybean. *Pest Manag Sci*. 2014;70(12):1831–1839.

18. Burr B. Recombinant inbreds for molecular mapping in maize: theoretical and practical considerations. *Trends Genet.* 1991;7(2):55–60.
19. Ananiev EV, Wu C, Chamberlin MA, et al. Artificial chromosome formation in maize (*Zea mays* L.). *Chromosoma.* 2009;118(2):157–177.
20. Cooper M, Gho C, Leafgren R, Tang T, Messina C. Breeding drought-tolerant maize hybrids for the US corn-belt: discovery to product. *J Exp Bot.* 2014;65(21):6191–6204.
21. Shi J, Drummond BJ, Wang H, Archibald RL, Habben JE. Maize and Arabidopsis ARGOS proteins interact with ethylene receptor signaling complex, supporting a regulatory role for ARGOS in ethylene signal transduction. *Plant Physiol.* 2016;171(4): 2783–2797.
22. de Pater BS, van der Mark F, Rueb S, et al. The promoter of the rice gene GOS2 is active in various different monocot tissues and binds rice nuclear factor ASF-1. *Plant J.* 1992;2(6):837–844.
23. Negrotto D, Jolley M, Beer S, Wenck AR, Hansen G. The use of phosphomannose-isomerase as a selectable marker to recover transgenic maize plants (*Zea mays* L.) via agrobacterium transformation. *Plant Cell Rep.* 2000;19(8):798–803.
24. Que Q, Chilton MD, de Fontes CM, et al. Trait stacking in transgenic crops: challenges and opportunities. *GM Crops.* 2010;1(4):220–229.
25. Eszterhas SK, Bouhassira EE, Martin DIK, Fiering S. Transcriptional interference by independently regulated genes occurs in any relative arrangement of the genes and is influenced by chromosomal integration position. *Mol Cell Biol.* 2002;22(2):469–479.
26. Li Z, Moon BP, Xing A, et al. Stacking multiple transgenes at a selected genomic site via repeated recombinase-mediated DNA cassette exchanges. *Plant Physiol.* 2010;154(2): 622–631.
27. Li Z, Liu ZB, Xing A, et al. Cas9-guide RNA directed genome editing in soybean. *Plant Physiol.* 2015;169(2):960–970.
28. Sebastian SAFG, Ulrich JF, Forney DR, Chaleff RS. Semidominant soybean mutation for resistance to sulfonylurea herbicides. *Crop Sci.* 1989;29:1403–1408.
29. Li Z, Xing A, Moon BP, McCardell RP, Mills K, Falco SC. Site-specific integration of transgenes in soybean via recombinase-mediated DNA cassette exchange. *Plant Physiol.* 2009;151(3):1087–1095.
30. Karvelis T, Gasiunas G, Young J, et al. Rapid characterization of CRISPR-Cas9 protospacer adjacent motif sequence elements. *Genome Biol.* 2015;16(1):253.
31. Zhang Y, Liang Z, Zong Y, et al. Efficient and transgene-free genome editing in wheat through transient expression of CRISPR/Cas9 DNA or RNA. *Nat Commun.* 2016;7:1–8.

CHAPTER THREE

Use of Zinc-Finger Nucleases for Crop Improvement

John P. Davies[1], Sandeep Kumar, Lakshmi Sastry-Dent
Dow AgroSciences, Indianapolis, IN, United States
[1]Corresponding author: e-mail address: jdavies3@dow.com

Contents

Abstract

Over the past two decades, new technologies enabling targeted modification of plant genomes have been developed. Among these are zinc-finger nucleases (ZFNs) which are composed of engineered zinc-finger DNA-binding domains fused with a nuclease, generally the *FokI* nuclease. The zinc-finger domains are composed of a series of four to six 30 amino acid domains that can bind to trinucleotide sequences giving the entire DNA-binding domain specificity to 12–18 nucleotides. Since the *FokI* nuclease functions as a dimer, pairs of zinc-finger domains are designed to bind upstream and downstream of the cut site which increases the specificity of the complete ZFN to 24–36 nucleotides. The ability of these engineered nucleases to create targeted double-stranded breaks at designated locations throughout the genome has enabled precise deletion, addition, and editing of genes. These techniques are being used to create new genetic variation by deleting or editing endogenous gene sequences and enhancing the efficiency of transgenic product development through targeted insertion of transgenes to specific genomic locations and to sequentially add and/or delete transgenes from existing transgenic events.

Progress in Molecular Biology and Translational Science, Volume 149
ISSN 1877-1173
http://dx.doi.org/10.1016/bs.pmbts.2017.03.006

1. INTRODUCTION

As the global population continues to rise, there is an increasing need to produce more food with limited natural resources. Historically, the productivity in the major crops has increased 1%–2% per year, but more rapid increases in productivity are projected to be required to meet future needs.[1] Technological advancements in both plant breeding and crop management, including new crop protection products, have driven productivity increases in the past[2] and will continue to do so in the future. Underpinning past advancements in plant breeding were technologies such as the production of hybrid maize in the 1930s, increasing genetic variation through chemical and radiation mutagenesis, doubled haploid technology, and exploiting cell biology and tissue culture for micropropagation to capture highly productive lines. Advances in molecular biology and genomics have led to transgenic insect resistance and herbicide tolerance traits, marker-assisted breeding, and stacked trait products. Precise editing of plant genomes is a revolutionary new approach that allows specific modifications in plant genomes to accelerate discovery and development of new agricultural products. Genome-editing technologies are enhancing options for new product development, and they hold the potential to contribute to increases in crop productivity in the future.

In the past 20 years, several genome-editing technologies, including meganucleases, zinc-finger nucleases (ZFNs), transcription activator-like effector nucleases (TALENs), and the clustered regularly interspaced short palindromic repeats (CRISPR)/Cas system, have been developed for targeted plant genome modification. In general, these approaches use site-directed nucleases (SDNs) composed of DNA-binding domains linked to a nuclease domain. The DNA-binding domain provides sequence specificity, and the nuclease domain introduces double-stranded DNA breaks at the desired sequence. The double-stranded breaks (DSBs) generated by these reagents are repaired by host repair pathways such as nonhomologous end joining (NHEJ) and homology-directed repair (HDR). NHEJ is an imprecise DNA repair pathway that frequently results in insertions or deletions at the DSB site; repair of an SDN-induced break by NHEJ can result in mutagenesis. HDR is a precise DSB repair pathway that requires sequences homologous (almost identical) to those flanking the DSB site.[3,4] Exogenous nucleotide sequences (transgenes or native genomic sequences with intended mutations) can be inserted at the SDN-induced DSB sites by either

NHEJ or HDR, leading to targeted gene insertion or editing. Thus, SDNs and natural DNA repair mechanisms have been used to facilitate a range of targeted genome modifications including insertions, deletions, and edits in plants and animals.

Meganucleases (also referred to as LAGLIDADG homing endonucleases[5]) were the first tools used for plant genome editing.[6] These naturally occurring proteins function as homodimers comprised of two identical subunits each 160–200 amino acid residues in size, but they are also active as a single-fused peptide composed of two tandem repeat monomers joined together by a linker sequence.[7] Meganucleases typically bind to target sites of 20–30 bp; unlike ZFNs, TALENS, and the CRISPR/Cas system the DNA-binding domains of most meganucleases overlap with their catalytic domains[8] making them challenging to reengineer for editing new sites.[9,10]

ZFNs are engineered SDNs composed of a DNA-binding domain (zinc finger) and a nuclease domain. Each designed zinc finger (ZF) recognizes a specific 3-bp DNA sequence. Typically four to six individual ZFs are linked together to recognize a unique DNA sequence (12–18 bp). Since the *Fok*I nuclease functions as a dimer to cleave double-strand DNA, two ZF proteins are required to target the desired DNA sequence (24–36 nucleotides). ZFNs have been used to modify the genomes of many eukaryote cells or organisms including human,[11–17] plant,[18] *Drosophila melanogaster*,[19] *Caenorhabditis elegans*,[20] zebrafish,[19] mice,[21] pig,[22] and frog.[23]

TALENs are engineered SDNs in which a transcription activator-like effector (TALE) domain is fused to the catalytic domain of the *Fok*I endonuclease.[24] TALE binding to DNA is mediated by a central region of the TALE protein that contains as many as 30 tandem repeats of a 33–35 amino acid sequence motif[25]; each repeat binds a single base pair in the target DNA.[26]

RNA-guided genome editing using the CRISPR/Cas system has been recently developed as a genome-editing tool in plants.[27] The CRISPR genomic locus of *Streptococcus pyogenes* encodes the Cas9 (CRISPR-associated) endonuclease, which forms a complex with two short RNA molecules known as the CRISPR RNA (crRNA) and transactivating crRNA (tracrRNA). An engineered dual-tracrRNA:crRNA functions as a single-guide RNA (sgRNA) chimera and directs Cas9-mediated sequence-specific DNA cleavage.[28] The specificity of the nuclease is defined by the sgRNA; the Cas9 protein does not require any reengineering.

In this chapter, we review the development and use of genome-editing technologies based on ZFNs and how they are being used to develop the next generation of trait products.

2. WHY USE ZFNs?

The long track record of performance, specificity, and intellectual property rights makes ZFNs an attractive genome-editing tool for biotechnology product development. ZFNs have minimal target sequence restrictions and can be designed for most genes and regions of genomes of interest. Dow AgroSciences's genome-editing platform, branded as the EXZACT™ Precision Technology, is based on ZFNs.[29–37] Dow AgroSciences acquired an exclusive license from Sangamo BioSciences to use ZFNs in plants and has validated the technology in monocots and dicots. EXZACT™ technology has been used to delete, add, and edit the genomes of crop species. To date, Dow AgroSciences and its collaborators have implemented the EXZACT™ Precision Technology to modify (DNA addition, deletion, or editing) the genomes of more than 15 species targeting more than 150 genomic sites. Effort has also focused on building plant cell-based rapid testing systems for assessing ZFN functionality[38] and high-throughput analytical methods for identifying targeted plants.[39] Optimal genomic loci for targeted gene insertion have been identified in crops of interest,[40,41] and trait landing pads (TLPs) have been used to stack traits.[30]

ZFNs have certain advantages compared with CRISPR technology. The ability of ZFNs to bind and cleave at most locations with high fidelity is contrasted with the CRISPR/Cas system, which requires the presence of a protospacer adjacent motif (PAM) sequence at the target, which limits targets for genome editing.[28,42] In addition, several reports have described the off-target effects of CRISPRs, possibly due to the short 22-bp gRNA/PAM sequence that drives target specificity[43–45]; this feature may limit the use of the technology in complex polyploid plant genomes containing paralogous genes. Off-target effects appear to be minimal with high-quality ZFN designs built by Sangamo BioSciences or Sigma Aldrich (licensed ZFN provider) as demonstrated by specificity and efficacy of ZFNs in several crops including corn,[36] canola, and wheat (unpublished results). In addition, *Fok*I variants requiring heterodimerization have been developed that further enhance sequence specificity and reduce off-target cleavage.[46] A CCR5-specific ZFN was evaluated in an HIV1 clinical trial and deemed to be safe, which further illustrates to the specificity of these targeting reagents.[47]

Finally, Dow AgroSciences has exclusively licensed IP rights to the EXZACT™ Precision Technology from Sangamo BioSciences who secured its rights to ZFNs through licensing and developing patent and other proprietary rights. Conversely, the IP rights around CRISPRs are more complex and may continue to be under dispute. The first patent applications for the CRISPR–Cas system were filed in 2012, and already the CRISPR–Cas technology is enmeshed in an ongoing patent interference between The Broad Institute, Inc. and The Regents of the University of California. Though the Patent Trial and Appeal Board ruled in favor of the Broad Institute, it is likely that the legal battle over CRISPRs will continue to make it difficult to have a clear path to commercializing a product.[48,49]

3. PLANT GENOME EDITING USING ZFNs

3.1 Targeted Mutagenesis

The ability to modify specific gene sequences to create loss-of-function variants is a powerful technique for the investigation of gene function as well as the development of new products. Designed nucleases have become a potent tool for creating mutations at numerous sites in many species.[50] Examples of ZFN-induced mutations generated through DSBs and NHEJ include the *Arabidopsis* mutations in ABA-insensitive 4 (*ABI4*),[51] alcohol dehydrogenase 1 (*ADH1*), and transparent testa-4 (*TT4*)[52]; maize mutations in *IPK1* to create low phytate lines[36]; tomato mutations in the mitochondrial malate dehydrogenase gene[37]; and soybean mutations in the *DCL4a* and *DCL4b* genes.[53] These examples demonstrate the potential for creating valuable mutations in important crop species. In addition to supporting basic plant biology research, this technology has implications for applied plant breeding and crop improvement through the creation of novel traits by deletion of genes that are deleterious or that repress agronomically beneficial traits.

The ability to delete selectable marker sequences from the genome of transgenic plants is also an important biotechnological tool for transgenic trait deployment. Although selectable markers are critical for creation and regeneration of transgenic plants, they are often not needed for trait products.[54] In addition, current trait development may involve sequential addition and/or deletion of selectable markers.[55] Removal of selectable markers and other transgenic sequences has been demonstrated using both site-specific recombinases (SSRs)[35,56–58] and engineered nucleases.[59–61] While gene deletion via SSRs, such as Cre/Lox, is limited to artificially created transgenic locations that contain preintegrated recombination sites, ZFN-mediated gene deletion can occur at virtually any site in the genome.[62]

3.2 Targeted Gene Addition: Trait Targeting

In order to develop effective products, transgenes need to be integrated into genomic sites that support the desired level of stable expression over many generations. Random integration of the transgenes into the host plant genome can have detrimental effects and require extensive screening and characterization of several transgenic events to identify transgenic plants that are substantially equivalent to nontransgenic plants. Integration at random loci may lead to variable transgene expression caused by position effects and mutations in genic regions.[63] ZFN-mediated gene targeting (GT) can mitigate these effects by delivering new transgenes to investigator-defined genomic locations. Regions that support adequate transgene expression without perturbing endogenous gene structure or function are termed *safe harbors*.[64] In the human genome, "safe harbors" are located outside of transcription units and ultraconserved regions, ~50 kb from the 5′ end of any gene, and ~300 kb from any cancer-related or microRNA gene.[65] In rice, some potential "safe harbor" loci have been characterized using the reporter gene β-glucuronidase.[66] The integration of trait-conferring transgenes into a "safe harbor" could aid in development of transgenic products by providing a location for predictable transgene expression, so fewer events would need to be generated and characterized. Additionally, such "safe harbor" loci can be reused for multiple product development campaigns, resulting in savings in cost, time, and resources and streamlining processes. Dow AgroSciences has developed specific criteria for identifying optimal loci in plant genomes and demonstrated DNA insertion via ZFNs at some of these genomic loci. These optimal loci were identified using novel genomic features that had not been previously described or considered.[40,41] Transgene insertion loci in events generated by random transformation may also be used for retargeting provided they meet some or all of the optimal loci attributes; one such locus in corn, Event 32, has been reproducibly targeted in wild-type Hi-II and B104 genotypes using ZFNs at frequencies of 1%–3%.[29]

Targeting genes to specific locations in plant genomes was first demonstrated by repairing defective reporter genes in transgenic plants.[32,67,68] Subsequently, a donor transgene containing an herbicide tolerance gene without a promoter was used to target and trap an endogenous promoter sequence, which provided preferential selection of targeted cells.[36] Targeted insertion of complete transgenes into native loci using ZFNs has also been demonstrated.[36] However, random insertions of the transgene elsewhere in the genome may also occur, and when it does, it must be eliminated through breeding. A positive and negative selection strategy may be useful in deploying GT technology in plants. This approach uses selection markers to enrich for cells carrying an insertion, while eliminating events containing random insertions through negative selection.[69,70] These GT strategies provide powerful tools for a single round of targeting, but do not offer sequential GT for multigenerational trait stacking that is likely to be needed in the future.

3.2.1 Sequential and Modular GT

Modern agriculture requires a trait generation platform to meet complex product needs. The current transgenic trait development process requires stacking of multiple genes to pyramid an increased number of transgenes that have multiple modes of action for each trait. There has been a dramatic increase in the number of transgenic crop offerings with stacked traits.[71] One example, SmartStax™, a maize trait package offered by Dow AgroSciences and Monsanto, consists of eight transgenes for herbicide tolerance and insect resistance composed of four separate transgenic insertion events. New insect resistance or herbicide tolerance transgenes that may be stacked with these will further increase the number of loci as well as the complexity of pyramiding these transgenes. Currently, these stacks are generated through conventional breeding techniques, which require crossing plants containing several randomly integrated and independently segregating transgenic events with high-yielding elite lines. The events to be incorporated are usually in genetic backgrounds that differ from elite inbreds, hybrids, or varieties into which they are introgressed and the DNA of the donor line containing the trait must be replaced with that of the elite line through recombination to prevent linkage drag and other yield limiting characteristics.[71,72] The overall goal of trait stacking is to produce a finished line that contains traits that are consistently expressed and that yield the same as or better than the unconverted line. Because seed companies want to minimize the risk of not recovering an equivalent line, multiple versions of introgression are generated and evaluated simultaneously.[72] In addition, since seed companies sell many different elite hybrids and varieties and release new ones every year, the process of trait introgression and stacking is continuous. For example, Dow AgroSciences currently offers over 50 SmartStax™ hybrids. These factors make the introgression and stacking of traits into elite germplasm complex and expensive.

The next-generation biotech trait products are likely to be developed by adding new transgenes to existing ones or replacing the old transgenes with new ones.[55] Engineered nucleases have the capability of enabling this flexibility. Engineered nucleases have been used to target integration of new transgenes to a previously created transgenic locus.[30,73,74] For example, the single-chain I-CreI-based engineered homing endonuclease has been used to insert two herbicide tolerance genes near a preexisting transgenic insect control locus in cotton where the targeted sequence insertion events were recovered at a frequency of 1.8% of the putative transgenic events on the selection media.[10] Ainley *et al.* used ZFNs to target the herbicide tolerance gene *aad1* to a locus already containing the *pat* herbicide tolerance gene and an integrated TLP.[30] In these cases, many plants had to be screened to find the targeted events within the background of randomly inserted events. A system that preferentially inserts transgenes at a known location without random integration is desirable for routine application and sequential targeting of new transgenes for trait development.

A new selection-based sequential GT method was recently developed using ZFN-mediated DSBs.[33] This method relies on swapping different fully functional selection markers every cycle of targeting. The selection marker in the target locus is driven by a promoter that contains a unique intron, while the donor molecule contains a different selection marker with the same intron but without any promoter. A ZFN-mediated DSB in the intron and homology between target and donor leads to the exchange of the selection marker cassettes and provides positive selection for site-specific integration of the donor. This nuclease-mediated cassette exchange (NMCE) method could be used over multiple generations to add new genes and remove or replace old ones.

3.2.2 In Planta Targeting and Intragenomic Recombination

In most examples of GT to date, the need for a robust plant transformation and tissue culture system has been an essential component for genome-editing protocols developed for crop plants. Both donor and nuclease-encoding DNA molecules are typically delivered to plant cells through direct (e.g., particle bombardment or WHISKERS™) or *Agrobacterium*-mediated delivery systems, and targeted events are generated in the small number of cells that receive and contain both the donor molecule and the nuclease. The frequency of targeting is therefore dependent on the efficiencies of plant transformation and regeneration. The efficiencies of both transformation and regeneration are relatively low and limit deployment of GT technologies in plant breeding programs and commercial agriculture industry.

In planta GT may minimize the reliance of GT on transformation and tissue culture processes. This targeting method requires a very small number of primary transgenic plants with stable integrations of the donor DNA and a gene encoding the nuclease. The randomly integrated donor transgene is flanked by binding sites specific to the nuclease being expressed. DSBs at these sites lead to excision of the donor and enhance the potential of it to recombine with homologous sequences in the target locus. Unlike conventional transformation-mediated GT methods, the advantage of in planta targeting is that the donor and nuclease are in every cell so that targeting can occur in a large number of cells throughout the life cycle of the plant.

Although in planta GT was proposed over a decade ago, the first practical demonstration was only recently reported. Fauser *et al.*[75] described using a homing endonuclease gene *I-SceI* for in vivo liberation of a donor sequence and its precise integration into a target site in *Arabidopsis*. The authors reported ~1% GT based on the number of seeds analyzed in the F_2 generation.[75] This was followed by work in maize showing *I-SceI*-mediated GT[76] where targeted plants were identified by subjecting F_2 immature embryos to callus induction and regeneration on selective medium. Using ZFN, a similar in planta targeting system was reported in maize via intragenomic homologous recombination and NMCE. In this case, up to 3.3% of the plants regenerated on selection indicated insertion of the donor at the target locus.[77] These initial reports demonstrating in vivo targeting in plants have created highly promising prospects for GT in applied plant breeding and commercial agriculture. The technology is still at early stage and the need for a tissue culture process restricts its application to a limited number of genotypes. However, future GT systems that target germline or reproductive cells hold potential promise not only alleviate the tissue culture step but also extend application of this approach beyond transgenes to native traits across multiple crops, leading to future designer crops.

4. TARGETED GENE EDITING

In most eukaryotic species, gene replacement, the process of replacing one gene for another, is an inefficient technology. The use of engineered nucleases has greatly enhanced the efficiency of this process. Like GT, the underlying technology relies on creating double-stranded DNA breaks at a specific site, providing a donor template that is suitable for homologous recombination and relying on the host HDR processes to catalyze the insertion of the donor template into the site of the DSB.

Gene replacement using ZFN-induced double-stranded DNA breaks was pioneered in *Drosophila*,[78] which showed that ZFN-induced cuts at the *y* locus in germline cells facilitated the replacement of a wild-type form of the *y* gene with a mutant form when the mutated version was provided as a donor sequence. The replacement was most efficient when this donor sequence was provided as a linear molecule where gene replacement rates of ~2% were achieved. Since then, ZFN-mediated gene replacement has been demonstrated in many eukaryotic systems including mice,[79] rats,[80] humans[12,81–84], and plants.[32,36,85]

The ability to replace one version of a gene for another introduces new possibilities in engineering plants. DSB-mediated gene replacement has already been used to generate herbicide tolerant plants by replacing wild-type versions of the acetolactate synthase genes with mutated ones,[31,86,87] and high oil soybeans by replacing diacylglycerol acyltransferase with a mutated version.[88] The ability to make these precise gene edits will enable the modification of pathways that enhance the biochemistry of food and medicinal plants in ways that add value to the grower and consumer.

5. ENHANCING THE EFFICIENCY OF GENOME EDITING

The efficiency of gene replacement by HDR is known to be affected by several factors including the length and sequence identity of the homology arms and the topology of the sequence. Homology arms need to be 200–300 bp, but longer homology arms can increase the efficiency. Single-stranded oligonucleotides have been shown to be targeted much more efficiently than double-stranded fragments.[89]

The efficiency of HDR is also influenced by recombination mechanisms operating in the cell. HDR uses the endogenous repair mechanisms that operate only during the S and G2 phases of the cell cycle. It also competes with NHEJ, which functions throughout the cell cycle.[90] Recently, mutations in components of the NHEJ pathway have been shown to increase the efficiency of HDR-mediated gene replacement relative to that in non-mutated organisms. In *Drosophila* and mice as well as in *Arabidopsis*, GT in mutants deficient for DNA ligase IV, a component of the NHEJ-repair pathway, was up to 19 times more efficient than in nonmutant organisms.[89,91–93]

Using nucleases that cause double-stranded DNA breaks, the efficiency of the gene editing by HDR has been shown to be between 0.5% and 10% of the transformed plants.[30,94] SSRs and transposases may be alternatives to site-specific nucleases for catalyzing gene replacement. Both SSRs and transposases directly catalyze the cleavage, strand exchange, and ligation of homologous DNA sequences and thus do not rely on endogenous homologous recombination machinery or provoke error-prone DNA repair processes.[95,96] SSRs such as Cre-*lox* and FLP-*FRT* have been used to catalyze recombination between donors and targets.[97] However, these enzymes carry out a reversible recombination reaction so that fragments are recombined in and out by the same enzyme. Strategies have been developed that stabilize the integration of DNA using mutant *lox* sites.[97,98]

The known SSRs have strict sequence specificity. In order to make these enzymes more useful for gene replacement, the endogenous sequence specificity needs to be overcome and mechanisms to direct the enzymes to endogenous sequences in target organisms need to be developed. Recently, engineered recombinases with relaxed sequence specificity have been fused with ZF DNA-binding domains and shown to achieve rates of site-specific integration similar to engineered nuclease systems.[99] A similar approach using "recCas9," an RNA-programed small serine recombinase, has been shown to function in human cells with an efficiency as high as 32% and the piggyBAC transposase has also been fused with a TALE DNA-binding domain and shown to direct sequence-specific insertion.[100] The use of modified SSRs and transposases to catalyze sequence-specific insertion may lead to enhancements in the efficiency and fidelity of the integration process.

6. REGULATORY FRAMEWORK

Globally, regulatory oversight of biotechnology-derived products is expensive and time consuming, with costs in many millions of dollars and requiring over 6 years for regulatory approval and deregulation.[101] Genome-editing tools can introduce mutations that can be achieved via traditional mutagenesis with far more precision and in less time. Such genome-edited crops can result in products that are identical to those generated using conventional methods. Based on these facts, USDA APHIS has concluded that products developed by deleting native gene sequences using genome-editing tools fall outside the scope of their regulation. For example, low phytate corn (developed by Dow AgroSciences) using EXZACT™ Precision Technology (USDA APHIS, 2014), drought-resistant corn (developed by DuPont Pioneer) created using CRISPR, browning-resistant mushrooms (developed by Pennsylvania State University), and the bruise-resistant, nonbrowning potatoes created using TALENs (developed by Calyxt)[102] do not fall under the scope of their regulation. It should be noted that all of these products were generated using the genome-editing technologies to delete the genomic nucleotides and were deemed to not contain plant pest DNA. Genome-edited canola developed using an oligonucleotide method was confirmed recently to be non-GMO by German authorities due to its similarity to products developed via conventional methods.[103]

There is urgency across the globe to establish a regulatory framework for genome-edited crops, so society can leverage the benefits of these crops. The scientific community is proposing product-based regulation rather than technology-based regulation for edited crops due to the precise nature of genome modifications.[103,104] Changes and harmonization of global regulatory framework around genome-edited crops are imperative for future progress of the field.

7. CONCLUDING STATEMENTS

Genome editing in plants has been demonstrated with ZFNs and other targeting reagents. Application of genome-editing technologies for functional genomics, trait discovery and development, and establishment of a regulatory framework for products developed using editing technologies are critical for fully exploiting these methods for food productivity to benefit the global population.

REFERENCES

1. Ray DK, Mueller ND, West PC, Foley JA. Yield trends are insufficient to double global crop production by 2050. *PLoS One*. 2013;8(6):e66428.
2. Duvick DN. The contribution of breeding to yield advances in maize (Zea mays L.). *Adv Agron*. 2005;86:83–145.
3. Lieberman-Lazarovich M, Levy AA. Homologous recombination in plants: an anti-review. *Methods Mol Biol*. 2011;701:51–65.
4. Waterworth WM, Drury GE, Bray CM, West CE. Repairing breaks in the plant genome: the importance of keeping it together. *New Phytol*. 2011;192:805–822.
5. Daboussi F, Stoddard TJ, Zhang F. Engineering meganuclease for precise plant genome modification. In: Zhang F, Puchta H, Thomson JG, eds. *Advances in New Technology for Targeted Modification of Plant Genomes*. New York: Springer; 2015:21–38.
6. Puchta H, Dujon B, Hohn B. Homologous recombination in plant cells is enhanced by in vivo induction of double strand breaks into DNA by a site-specific endonuclease. *Nucleic Acids Res*. 1993;21(22):5034–5040.

7. Stoddard BL. Homing endonucleases: from microbial genetic invaders to reagents for targeted DNA modification. *Structure*. 2011;19(1):7–15.
8. Taylor GK, Petrucci LH, Lambert AR, Baxter SK, Jarjour J, Stoddard BL. LAHEDES: the LAGLIDADG homing endonuclease database and engineering server. *Nucleic Acids Res*. 2012;40(Web Server issue):W110–W116.
9. Puchta H, Fauser F. Synthetic nucleases for genome engineering in plants: prospects for a bright future. *Plant J*. 2014;78(5):727–741.
10. D'Halluin K, Ruiter R. Directed genome engineering for genome optimization. *Int J Dev Biol*. 2013;57(6–8):621–627.
11. Lombardo A, Genovese P, Beausejour CM, et al. Gene editing in human stem cells using zinc finger nucleases and integrase-defective lentiviral vector delivery. *Nat Biotechnol*. 2007;25(11):1298–1306.
12. Urnov FD, Miller JC, Lee Y-L, et al. Highly efficient endogenous human gene correction using designed zinc-finger nucleases. *Nature*. 2005;435(7042):646–651.
13. Moehle EA, Rock JM, Lee Y-L, et al. Targeted gene addition into a specified location in the human genome using designed zinc finger nucleases. *Proc Natl Acad Sci USA*. 2007;104(9):3055–3060.
14. Perez EE, Wang J, Miller JC, et al. Establishment of HIV-1 resistance in CD4+ T cells by genome editing using zinc-finger nucleases. *Nat Biotechnol*. 2008;26(7): 808–816.
15. Sebastiano V, Maeder ML, Angstman JF, et al. In situ genetic correction of the sickle cell anemia mutation in human induced pluripotent stem cells using engineered zinc finger nucleases. *Stem Cells*. 2011;29(11):1717–1726.
16. Wilen CB, Wang J, Tilton JC, et al. Engineering HIV-resistant human CD4+ T cells with CXCR4-specific zinc-finger nucleases. *PLoS Pathog*. 2011;7(4):e1002020.
17. Provasi E, Genovese P, Lombardo A, et al. Editing T cell specificity towards leukemia by zinc finger nucleases and lentiviral gene transfer. *Nat Med*. 2012;18(5):807–815.
18. Petolino JF. Genome editing in plants via designed zinc finger nucleases. *In Vitro Cell Dev Biol Plant*. 2015;51(1):1–8.
19. Doyon Y, McCammon JM, Miller JC, et al. Heritable targeted gene disruption in zebrafish using designed zinc-finger nucleases. *Nat Biotechnol*. 2008;26(6):702–708
20. Morton J, Davis MW, Jorgensen EM, Carroll D. Induction and repair of zinc-finger nuclease-targeted double-strand breaks in Caenorhabditis elegans somatic cells. *Proc Natl Acad Sci USA*. 2006;103(44):16370–16375.
21. Osiak A, Radecke F, Guhl E, et al. Selection-independent generation of gene knockout mouse embryonic stem cells using zinc-finger nucleases. *PLoS One*. 2011;6(12): e28911.
22. Hauschild J, Petersen B, Santiago Y, et al. Efficient generation of a biallelic knockout in pigs using zinc-finger nucleases. *Proc Natl Acad Sci USA*. 2011;108(29):12013–12017.
23. Young JJ, Cherone JM, Doyon Y, et al. Efficient targeted gene disruption in the soma and germ line of the frog Xenopus tropicalis using engineered zinc-finger nucleases. *Proc Natl Acad Sci USA*. 2011;108(17):7052–7057.
24. Cermak T, Doyle EL, Christian M, et al. Efficient design and assembly of custom TALEN and other TAL effector-based constructs for DNA targeting. *Nucleic Acids Res*. 2011;39(12):e82.
25. Christian M, Cermak T, Doyle EL, et al. Targeting DNA double-strand breaks with TAL effector nucleases. *Genetics*. 2010;186(2):757–761.
26. Boch J, Scholze H, Schornack S, et al. Breaking the code of DNA binding specificity of TAL-type III effectors. *Science*. 2009;326(5959):1509–1512.
27. Shan Q, Wang Y, Li J, et al. Targeted genome modification of crop plants using a CRISPR-Cas system. *Nat Biotechnol*. 2013;31(8):686–688.

28. Jinek M, Chylinski K, Fonfara I, Hauer M, Doudna JA, Charpentier E. A programmable dual-RNA-guided DNA endonuclease in adaptive bacterial immunity. *Science*. 2012;337(6096):816–821.
29. Ainley WM, Bing JW, Corbin DH, et al. Precision gene targeting to a particular locus in maize. US patent application 14/105,563; 2013.
30. Ainley WM, Sastry-Dent L, Welter ME, et al. Trait stacking via targeted genome editing. *Plant Biotechnol J*. 2013;11(9):1126–1134.
31. Ainley WM, Guschin DY, Hayden M, et al. Methods and compositions for integration of an exogenous sequence within the genome of plants. US patent application US 14/243,553; 2014.
32. Cai CQ, Doyon Y, Ainley WM, et al. Targeted transgene integration in plant cells using designed zinc finger nucleases. *Plant Mol Biol*. 2009;69(6):699–709.
33. Kumar S, AlAbed D, Worden A, Novak S, Wu H, Ausmus C. A modular gene targeting system for sequential transgene stacking in plants. *J Biotechnol*. 2015;207: 12–20.
34. Kumar S, Barone P, Smith M. Gene targeting and transgene stacking using intra genomic homologous recombination in plants. *Plant Methods*. 2016;12(1):1–7.
35. Kumar S, Thompson WF. Simultaneous excision of two transgene flanking sequences and resolution of complex integration loci. *Plant Mol Biol*. 2009;69(1–2): 23–32.
36. Shukla VK, Doyon Y, Miller JC, et al. Precise genome modification in the crop species Zea mays using zinc-finger nucleases. *Nature*. 2009;459(7245):437–441.
37. Shukla V, Gupta M, Urnov F, Guschin D, De Both MJ, Bundock P. Targeted modification of malate dehydrogenase. US patent 9,523,098; 2013.
38. Cao Z, Novak S, Sastry-Dent L, Zhou N. High through-put analysis of transgene borders. WO2013078318A1; 2013.
39. Sastry-Dent L, Simpson M, Cao Z, Chen W, Zhou N, Webb SR. DNA detection methods for site specific nuclease activity. US patent 9,493,844; 2013.
40. Sastry-Dent L, Cao Z, Sriram S, Webb SR, Camper DL. Optimal maize loci. US patent application 14/531,739; 2014.
41. Sastry-Dent L, Cao Z, Sriram S, Webb SR, Camper DL. Optimal soybean loci. US patent application 14/531,732; 2014.
42. Zetsche B, Gootenberg JS, Abudayyeh OO, et al. Cpf1 is a single RNA-guided endonuclease of a class 2 CRISPR-Cas system. *Cell*. 2015;163(3):759–771.
43. Fu Y, Foden JA, Khayter C, et al. High-frequency off-target mutagenesis induced by CRISPR-Cas nucleases in human cells. *Nat Biotechnol*. 2013;31(9):822–826.
44. Hsu PD, Scott DA, Weinstein JA, et al. DNA targeting specificity of RNA-guided Cas9 nucleases. *Nat Biotechnol*. 2013;31(9):827–832.
45. Liang P, Xu Y, Zhang X, et al. CRISPR/Cas9-mediated gene editing in human tripronuclear zygotes. *Protein Cell*. 2015;6:363–372.
46. Miller JC, Holmes MC, Wang J, et al. An improved zinc-finger nuclease architecture for highly specific genome editing. *Nat Biotechnol*. 2007;25(7):778–785.
47. Tebas P, Stein D, Tang WW, et al. Gene editing of CCR5 in autologous CD4 T cells of persons infected with HIV. *N Engl J Med*. 2014;370(10):901–910.
48. Ledford H. Why the CRISPR patent verdict isn't the end of the story. *Nature News, Nature Publishing Group*. 2017.
49. Cross R. CRISPR patent dispute ends well for Broad Institute. Berkeley says: not so fast. *Chem Eng News*. 2017;95(8):7.
50. Carroll D. The development and use of zinc-finger nucleases. In: Cathomen T, Hirsch M, Porteus M, eds. *Genome Editing*. New York: Springer; 2016:15–28.
51. Osakabe K, Osakabe Y, Toki S. Site-directed mutagenesis in Arabidopsis using custom-designed zinc finger nucleases. *Proc Natl Acad Sci USA*. 2010;107(26): 12034–12039.

52. Zhang F, Maeder ML, Unger-Wallace E, et al. High frequency targeted mutagenesis in Arabidopsis thaliana using zinc finger nucleases. *Proc Natl Acad Sci USA*. 2010;107(26):12028–12033.
53. Curtin SJ, Anderson JE, Starker CG, et al. Targeted mutagenesis for functional analysis of gene duplication in legumes. In: Rose RJ, ed. *Legume Genomics: Methods and Protocols*. Totowa, NJ: Humana Press; 2013:25–42.
54. Rosellini D. Selectable markers and reporter genes: a well furnished toolbox for plant Science and genetic engineering. *Crit Rev Plant Sci*. 2012;31(5):401–453.
55. Kumar S, Barone P, Smith M. Gene targeting and transgene stacking using intra genomic homologous recombination in plants. *Plant Methods*. 2016;12:11.
56. Zhang W, Subbarao S, Addae P, et al. Cre/lox-mediated marker gene excision in transgenic maize (Zea mays L.) plants. *Theor Appl Genet*. 2003;107(7):1157–1168.
57. Srivastava V, Anderson OD, Ow DW. Single-copy transgenic wheat generated through the resolution of complex integration patterns. *Proc Natl Acad Sci USA*. 1999;96(20):11117–11121.
58. Nandy S, Zhao S, Pathak BP, Manoharan M, Srivastava V. Gene stacking in plant cell using recombinases for gene integration and nucleases for marker gene deletion. *BMC Biotechnol*. 2015;15:93.
59. Siebert R, Puchta H. Efficient repair of genomic double-strand breaks by homologous recombination between directly repeated sequences in the plant genome. *Plant Cell*. 2002;14:1121–1131.
60. Antunes MS, Smith JJ, Jantz D, Medford JI. Targeted DNA excision in Arabidopsis by a re-engineered homing endonuclease. *BMC Biotechnol*. 2012;12(1):86.
61. Petolino JF, Worden A, Curlee K, Connell J, Strange Moynahan TL, Larsen C. Zinc finger nuclease-mediated transgene deletion. *Plant Mol Biol*. 2010;73:617–628.
62. Petolino JF, Kumar S. Transgenic trait deployment using designed nucleases. *Plant Biotechnol J*. 2016;14(2):503–509.
63. Matzke AJ, Matzke MA. Position effects and epigenetic silencing of plant transgenes. *Curr Opin Plant Biol*. 1998;1(2):142–148.
64. Papapetrou EP, Lee G, Malani N, et al. Genomic safe harbors permit high beta-globin transgene expression in thalassemia induced pluripotent stem cells. *Nat Biotechnol*. 2011;29(1):73–78.
65. Bejerano G, Pheasant M, Makunin I, et al. Ultraconserved elements in the human genome. *Science*. 2004;304(5675):1321–1325.
66. Cantos C, Francisco P, Trijatmiko KR, Slamet-Loedin I, Chadha-Mohanty PK. Identification of "safe harbor" loci in indica rice genome by harnessing the property of zinc-finger nucleases to induce DNA damage and repair. *Front Plant Sci*. 2014;5:302.
67. Puchta H, Dujon B, Hohn B. Two different but related mechanisms are used in plants for the repair of genomic double-strand breaks by homologous recombination. *Proc Natl Acad Sci USA*. 1996;93:5055–5060.
68. Wright DA, Townsend JA, Winfrey RJ, et al. High-frequency homologous recombination in plants mediated by zinc-finger nucleases. *Plant J*. 2005;44(4):693–705.
69. Mansour SL, Thomas KR, Capecchi MR. Disruption of the proto-oncogene int-2 in mouse embryo-derived stem cells: a general strategy for targeting mutations to non-selectable genes. *Nature*. 1988;336(6197):348–352.
70. Shimatani Z, Nishizawa-Yokoi A, Endo M, Toki S, Terada R. Positive-negative-selection-mediated gene targeting in rice. *Front Plant Sci*. 2014;5:748.
71. Que Q, Chilton M-DM, de Fontes CM, et al. Trait stacking in transgenic crops: challenges and opportunities. *GM Crops*. 2010;1(4):220–229.
72. Peng T, Sun X, Mumm RH. Optimized breeding strategies for multiple trait integration: II. Process efficiency in event pyramiding and trait fixation. *Mol Breed*. 2014;33(1):105–115.

73. D'Halluin K, Vanderstraeten C, Stals E, Cornelissen M, Ruiter R. Homologous recombination: a basis for targeted genome optimization in crop species such as maize. *Plant Biotechnol J.* 2008;6(1):93–102.
74. D'Halluin K, Vanderstraeten C, Van Hulle J, et al. Targeted molecular trait stacking in cotton through targeted double-strand break induction. *Plant Biotechnol J.* 2013;11(8):933–941.
75. Fauser F, Roth N, Pacher M, et al. In planta gene targeting. *Proc Natl Acad Sci USA.* 2012;109(19):7535–7540.
76. Ayar A, Wehrkamp-Richter S, Laffaire JB, et al. Gene targeting in maize by somatic ectopic recombination. *Plant Biotechnol J.* 2013;11(3):305–314.
77. Kumar S, Worden A, Novak S, Lee R, Petolino JF. A trait stacking system via intra-genomic homologous recombination. *Planta.* 2016;244(5):1157–1166.
78. Bibikova M, Beumer K, Trautman JK, Carroll D. Enhancing gene targeting with designed zinc finger nucleases. *Science.* 2003;300(5620):764.
79. Li H, Haurigot V, Doyon Y, et al. In vivo genome editing restores haemostasis in a mouse model of haemophilia. *Nature.* 2011;475(7355):217–221.
80. Cui X, Ji D, Fisher DA, Wu Y, Briner DM, Weinstein EJ. Targeted integration in rat and mouse embryos with zinc-finger nucleases. *Nat Biotechnol.* 2011;29(1):64–67.
81. Genovese P, Schiroli G, Escobar G, et al. Targeted genome editing in human repopulating haematopoietic stem cells. *Nature.* 2014;510(7504):235–240.
82. Yusa K, Rashid ST, Strick-Marchand H, et al. Targeted gene correction of [agr] 1-antitrypsin deficiency in induced pluripotent stem cells. *Nature.* 2011; 478(7369):391–394.
83. Zou J, Mali P, Huang X, Dowey SN, Cheng L. Site-specific gene correction of a point mutation in human iPS cells derived from an adult patient with sickle cell disease. *Blood.* 2011;118(17):4599–4608.
84. Soldner F, Laganière J, Cheng AW, et al. Generation of isogenic pluripotent stem cells differing exclusively at two early onset Parkinson point mutations. *Cell.* 2011; 146(2):318–331.
85. Townsend JA, Wright DA, Winfrey RJ, et al. High-frequency modification of plant genes using engineered zinc-finger nucleases. *Nature.* 2009;459(7245):442–445.
86. Svitashev S, Young JK, Schwartz C, Gao H, Falco SC, Cigan AM. Targeted mutagenesis, precise gene editing, and site-specific gene insertion in maize using Cas9 and guide RNA. *Plant Physiol.* 2015;169(2):931–945.
87. Li Z, Liu Z-B, Xing A, et al. Cas9-guide RNA directed genome editing in soybean. *Plant Physiol.* 2015;169(2):960–970.
88. Roesler K, Shen B, Bermudez E, et al. An improved variant of soybean type 1 diacylglycerol acyltransferase increases the oil content and decreases the soluble carbohydrate content of soybeans. *Plant Physiol.* 2016;171(2):878–893.
89. Beumer KJ, Trautman JK, Mukherjee K, Carroll D. Donor DNA utilization during gene targeting with zinc-finger nucleases. *G3 (Bethesda).* 2013;3(4):657–664.
90. Coluccio A, Miselli F, Lombardo A, et al. Targeted gene addition in human epithelial stem cells by zinc-finger nuclease-mediated homologous recombination. *Mol Ther.* 2013;21(9):1695–1704.
91. Maruyama T, Dougan SK, Truttmann MC, Bilate AM, Ingram JR, Ploegh HL. Increasing the efficiency of precise genome editing with CRISPR-Cas9 by inhibition of nonhomologous end joining. *Nat Biotechnol.* 2015;33(5):538–542.
92. Beumer KJ, Trautman JK, Bozas A, et al. Efficient gene targeting in Drosophila by direct embryo injection with zinc-finger nucleases. *Proc Natl Acad Sci USA.* 2008;105(50):19821–19826.
93. Qi Y, Zhang Y, Zhang F, et al. Increasing frequencies of site-specific mutagenesis and gene targeting in Arabidopsis by manipulating DNA repair pathways. *Genome Res.* 2013;23(3):547–554.

94. Hilscher J, Bürstmayr H, Stoger E. Targeted modification of plant genomes for precision crop breeding. *Biotechnol J*. 2016;12:1600173.
95. Chaikind B, Bessen JL, Thompson DB, Hu JH, Liu DR. A programmable Cas9-serine recombinase fusion protein that operates on DNA sequences in mammalian cells. *Nucleic Acids Res*. 2016;44:9758–9770.
96. Mitra R, Fain-Thornton J, Craig NL. piggyBac can bypass DNA synthesis during cut and paste transposition. *EMBO J*. 2008;27(7):1097–1109.
97. Srivastava V, Thomson J. Gene stacking by recombinases. *Plant Biotechnol J*. 2016;14(2):471–482.
98. Albert H, Dale EC, Lee E, Ow DW. Site-specific integration of DNA into wild-type and mutant lox sites placed in the plant genome. *Plant J*. 1995;7(4):649–659.
99. Gaj T, Sirk SJ, Tingle RD, Mercer AC, Wallen MC, Barbas III CF. Enhancing the specificity of recombinase-mediated genome engineering through dimer interface redesign. *J Am Chem*. 2014;136(13):5047–5056.
100. Owens JB, Mauro D, Stoytchev I, et al. Transcription activator like effector (TALE)-directed piggyBac transposition in human cells. *Nucleic Acids Res*. 2013;41:9197–9207.
101. Smart RD, Blum M, Wesseler J. Trends in approval times for genetically engineered crops in the United States and the European Union. *J Agric Econ*. 2016;68:182–198.
102. Waltz E. CRISPR-edited crops free to enter market, skip regulation. *Nat Biotechnol*. 2016;34(6):582.
103. Huang S, Weigel D, Beachy RN, Li J. A proposed regulatory framework for genome-edited crops. *Nat Genet*. 2016;48(2):109.
104. Sprink T, Eriksson D, Schiemann J, Hartung F. Regulatory hurdles for genome editing: process- vs. product-based approaches in different regulatory contexts. *Plant Cell Rep*. 2016;35:1493–1506.

CHAPTER FOUR

Gene Editing in Polyploid Crops: Wheat, Camelina, Canola, Potato, Cotton, Peanut, Sugar Cane, and Citrus

Donald P. Weeks[1]
University of Nebraska, Lincoln, NE, United States
[1]Corresponding author: e-mail address: dweeks1@unl.edu

Contents

Abstract

Polyploid crops make up a significant portion of the major food and fiber crops of the world and include wheat, potato, cotton, apple, peanut, citrus, and brassica oilseeds such as rape, canola, and Camelina. The presence of three sets of chromosomes in triploids, four sets in tetraploids, and six sets in hexaploids present significant challenges to conventional plant breeding and, potentially, to efficient use of rapidly emerging gene and genome-editing systems such as zinc finger nucleases, single-stranded oligonucleotides, TALE effector nucleases, and clustered regularly interspaced short palindromic repeats (CRISPR/Cas9). However, recent studies with each of these techniques in several polyploid crops have demonstrated facile editing of some or all of the genes targeted for modification on homeologous chromosomes. These modifications have allowed improvements in food nutrition, seed oil composition, disease resistance, weed protection, plant breeding procedures, and food safety. Plants and plant products exhibiting useful new traits created through gene editing but lacking foreign DNA may face reduced regulatory restrictions. Such plants can be obtained either by simply

Progress in Molecular Biology and Translational Science, Volume 149
ISSN 1877-1173
http://dx.doi.org/10.1016/bs.pmbts.2017.05.002

selecting for null segregants that have lost their editing transgenes during plant breeding or, even more attractively, by delivery of biodegradable Cas9/sgRNA ribonucleoprotein complexes (i.e., no DNA) into plant cells where they are expressed only transiently but allow for efficient gene editing—a system that has been recently demonstrated in at least two polyploid crops. Such systems that create precise mutations but leave no transgene footprint hold potential promise for assisting with the elimination or great diminution of regulatory processes that presently burden approvals of conventional transgenic crops.

1. INTRODUCTION

Gene editing using zinc finger nucleases (ZFNs), single-stranded oligo DNA nucleotides (ssODNs), TALE effector nucleases (TALENs), or CRISPR/Cas9/sgRNA technologies holds great promise for expediting plant science research and rapidly improving agricultural crops in the near future.[1–3] This chapter focuses on the use of these technologies with polyploid plants—ones that represent several of our major crop species, but whose more complex genetic composition make them often more challenging to manipulate.[4–7] Indeed, an immediate recognition for those wishing to edit genes in a polyploid plant is that instead of two copies (alleles) of any particular gene present in a diploid plant, there are four copies of the same gene in a tetraploid plant or six in a hexaploid plant. While this may have been a definite concern in the early days of gene editing, more recent studies outlined in this chapter provide ample evidence that for several polyploid plant species, rapid and efficient modification can be achieved for most, if not all, chromosomes in the multiple chromosome sets of polyploid plants. Examples are presented in which gene-editing technologies have been successfully applied to triploids (apple, citrus), tetraploids (pasta wheat, potato, cotton, apple, and peanut), hexaploids (Camelina, bread wheat), and octaploids (sugar cane) for both academic studies and potential commercial applications (Table 1).

2. GENE EDITING IN POLYPLOID CROPS

2.1 Wheat

2.1.1 Creation of Gene Knockouts and Gene Fragment Deletions in Hexaploid Wheat Using CRISPR/Cas9

In the first report of successful use of the CRISPR/Cas9 gene-editing system in bread wheat (*Triticum aestivum*) cells,[7a] two genes, inositol

Table 1 Polyploid Crops Successfully Mutated Using Gene Editing Tools

Crop	Edit Tool	Target Gene(s)	Trait	References
Wheat	Cas9/ sgRNA	PDS and inox	Chlorophyll syn	Upadhyay *et al.* [7a]
	TALEN and Cas9/ sgRNA	TaMLO	Powdery mildew resist	Wang *et al.*[8]
	Cas9/ sgRNA	*TaGASR7*	Grain length and weight	Zhang *et al.*[9]
	Cas9/ sgRNA and geminivirus	Ubiquitin	Gene knockout	Gil-Humanes *et al.*[10]
	Cas9/ sgRNA RNPs	*TaGASR7*	Grain length and weight	Liang *et al.*[11]
Camelina	Cas9/ sgRNA	*FAD2*	Seed oil composition (high oleic and low polyunsaturated FAs)	Jiang *et al* [12] and Morineau *et al.*[13]
Oil seed rape	ssODNs	*ALS*	Herbicide resistance	Gocal *et al.*[14]
Potato	TALEN	Endogenous constit. promoter	Herbicide resistance	Forsyth *et al.*[15]
	TALEN and geminivirus	*ALS*	Herbicide resistance	Butler *et al.*[16]
	TALEN	Vacuolar invertase	No reducing sugars and improved food safety	Clasen *et al.*[17]
	TALEN and Cas9/ sgRNA	Granule-bound starch synthase	Altered starch composition	Kusano *et al.*[18] and Andersson *et al.*[19]
	Cas9/ sgRNA	S-genes	*Phytophthora infestans* resistance	Schaart *et al.*[20]
Cotton	Cas9/ sgRNA	Viral and satellite DNAs	Cotton leaf curl disease	Iqbal *et al.*[21]

Continued

Table 1 Polyploid Crops Successfully Mutated Using Gene Editing Tools—cont'd

Crop	Edit Tool	Target Gene(s)	Trait	References
Peanut	Various	Alergen genes	Peanut alergens	Chandran *et al.*[22]
	Various	Aflatoxin genes	Peanut mycotoxins	Arias *et al.*[23]
Sugar cane	TALEN	Caffeic acid *O*-methyltransferase	Reduced lignin improved biofuel prod	Jung and Altpeter[24]
Citrus	Cas9/sgRNA	*CsLOB1*	Canker resistance	Jia *et al.*[25]
Apple	Cas9/sgRNA RNPs	*DIPM-1, DIPM-2,* and *DIPM-4* genes	Resistance to fire blight disease	Malnoy *et al.*[26]

oxygenase (*inox*) and phytoene desaturase (*pds*), were targeted. In each case, efficient production of insertions and deletions were observed in wheat cell suspension cultures with each of two sgRNA constructs used to target each gene. When the two sgRNA genes were placed together in a single expression cassette, the gene fragment between the two target sites was deleted. These studies demonstrated that gene editing in cells of an allohexaploid crop species was possible and led the way to an abundance of genome engineering projects of significant value.

2.1.2 Development of Bread Wheat Resistant to Powdery Mildew Infection Using TALEN and CRISPR/Cas9 Technologies

Successful targeting of all six homoeoalleles (in A, B, and D genomes) of a particular gene in an allohexaploid crop such as bread wheat for knock-out was originally viewed as a daunting task. However, Wang *et al.*[8] were able to show that TALEN-induced mutation of all three *TaMLO* gene homoeologs in the same plant could be achieved and could confer heritable broad-spectrum resistance to the fungal disease, powdery mildew. They also demonstrated modifications of the same genes using the CRISPR/Cas9 system. This was an important first demonstration that a crop with a complex allohexaploid genome could be efficiently engineered using recently developed gene-editing techniques.

2.1.3 Efficient and Transgene-Free Genome Editing in Tetraploid and Hexaploid Wheat Varieties Through Transient Expression of CRISPR/Cas9 DNA or RNA

The presence of foreign DNA in transgenic plants triggers regulatory concerns that can delay the entry of a new genetically engineered plant into the marketplace. In the case of plants carrying mutant traits created through the use of any of the recently gene-editing tools, foreign genes used in editing can be eliminated by simple genetic crosses (e.g., Ref. 26a) and such plants may be exempt for regulation by governmental bodies such as USDA (e.g., Refs. 27–29). As a more direct approach that immediately eliminates the possibility that foreign DNA will be incorporated into the genome of plants subjected to gene editing, Zhang *et al.*[9] developed techniques for use in wheat callus tissues in which exogenous DNAs and/or RNAs are only transiently present and do not persist in mutagenized callus cells or regenerated wheat plants. The procedure that delivers DNA encoding Cas9 (and sgRNA) is called "transiently expressed CRISPR/Cas9 DNA" (abbreviated as TECCDNA) and the procedure delivering only RNAs is abbreviated TECCRNA (for "transient expression of the in vitro transcripts (IVTs) of Cas9-coding sequence and guide RNA"). Amazingly, homozygous mutant plants of both hexaploid bread wheat (*Triticum aestivum* L., AABBDD, $2n=6x=42$) and tetraploid durum wheat (*Triticum turgidum* L. var. *durum*, AABB, $2n=4x=28$) were identified in T0 populations—and none of these, as expected, contained detectable transgenic material.

2.1.4 High Efficiency Gene Targeting in Hexaploid Wheat Using DNA Replicons and CRISPR/Cas9

Another innovative approach to efficiently produce crops that contain targeted gene mutations but lack integrated transgene sequences has been developed by the Voytas laboratory.[30] This laboratory developed single-stranded DNA virus-based constructs that are capable of replicating in dicotyledonous plant cells and that carry editing genes (e.g., ZFN, TALEN, or Cas9 and sgRNA genes). These transiently expressed viral replicon constructs are often considerably more efficient in creating targeted mutations than traditional editing genes that are permanently integrated into the plant genome. Importantly, the viral replicon can also be constructed to contain a gene fragment as donor template that can integrate into an endogenous gene by homologous recombination (HR) when the gene-editing molecules create a double-stranded DNA break (DSB) in that target gene. More recently, the Voytas laboratory in collaboration with international partners[10] has extended this technology to hexaploid wheat by using portions of the wheat dwarf virus as the base viral replicon for carrying the Cas9/sgRNA gene-editing system. Using this replicon, they achieved a 110-fold increase in expression of a reporter gene relative to a conventional, chromosomally integrated, Cas9/sgRNA gene-editing system. By using a strong gene promoter to drive expression of the Cas9 gene, they were able to edit an endogenous ubiquitin gene at rates 12 times that achieved using a nonviral delivery system. Importantly, they also demonstrated gene targeted integration by HR in all of the three homoeoalleles (A, B, and D) of the hexaploid wheat genome—and that this occurred with a frequency of ~1% in infected wheat cells.

2.1.5 Efficient DNA-Free Genome Editing of Bread Wheat Using CRISPR/Cas9 Ribonucleoprotein Complexes

Perhaps the ultimate means of avoiding the use of foreign DNAs to achieve Cas9/sgRNA-based gene editing is to deliver directly into living cells a Cas9/sgRNA ribonucleoprotein (RNP) complex preassembled in vitro using purified Cas9 produced in *Escherichia coli* and synthetic sgRNA. This was first achieved in animal cells[31] and has more recently been demonstrated in plant cells.[32,33,33a] This important technological advance has recently been extended to use for efficient gene editing in hexaploid wheat by the Gao laboratory and collaborators.[11] Impressively, the entire process from preparation of Cas9/sgRNA complexes, bombardment of immature wheat embryos (100–400/experiment), regeneration of plantlets, and analyses of mutant DNA sequences required only 7–9 weeks. Two separate genes in each of two wheat varieties were tested with similar results. The data showed that generally four to five independent mutants were obtained from 100 bombarded embryos. Importantly, unlike plant cells transformed with Cas9 and sgRNA genes in which noticeable off-target mutations could be detected by deep sequencing, no off-target modifications were found in mutant plants generated from Cas9/sgRNA RNP bombardment. This was attributed to the likely short lifetimes of the Cas9/sgRNA complexes within the cytoplasm and nucleus of cells into which the complexes were delivered. Because plants obtained by this method are completely transgene free and yet contain the precisely mutated gene desired, it may be possible that the regulatory hurdles for wheat and other crop plants improved using Cas9/sgRNA RNP complexes will be less formidable than those presently in place for classical GMOs.

2.2 Camelina, Canola, and Oilseed Rape

2.2.1 Improving Oil Quality in Camelina Seeds Using CRISPR/Cas9

The Brassicaceae oil seed crops rapeseed, canola, and Camelina are important sources of cooking and industrial oils. As with many other vegetable oils, the oils from these brassica species are relatively high in polyunsaturated fatty acids such as linoleic acid (18:2) and linolenic acid (18:3) and low in monounsaturated fatty acids such as oleic acid (18:1). In the case of Camelina, linoleic acid comprises ~16% of total fatty acids, linolenic acid ~35%, and oleic acid ~16%. To provide vegetable oil that is both healthier and better suited for many industrial purposes (including biofuels), it would be desirable to produce Camelina oil with a significant increase in oleic acid and marked decreases in linoleic acid and linolenic acid.

With this goal in mind, one group of researchers in the United States[12] and another in France[13] set out to use the CRISPR/Cas9 gene-editing system to target the *fatty acid desaturase 2* (*FAD2*) genes for knockout. In so doing, synthesis of both linoleic acid and linolenic acid should be blocked and their precursor, oleic acid, should increase. The allohexaploid nature of the Camelina genome means that a total of six *FAD2* genes (three independent pairs of *FAD2* genes on each of three pairs of homeologous chromosomes) need to be targeted for disruption. The US lab[12] took advantage of the fact that the *FAD2* genes of Camelina are highly homologous to the *FAD2* gene in the diploid, *Arabidopsis thaliana*. Thus, they were able to design sgRNAs that would target the identical sequences in both Camelia and Arabidopsis, but first test the ability of these constructs to knockout the two allelic copies of the *FAD2* gene in Arabidopsis—a much simpler and faster task in Arabidopsis than in Camelina. Indeed, homozygous knockout lines of Arabidopsis were produced by no later than the T2 generation and, as expected, these seeds contained nearly 60% oleic acid (18:1) and less than 8% polyunsaturated fatty acids (18:2 and 18:3). When used to knockout the *FAD2* genes in Camelina, these same Cas9/sgRNA gene constructs resulted in T4 generation seeds containing over 50% oleic acid and less than 15% polyunsaturated fatty acids. This group presented data indicating that none of their plants contained knockouts of all six *FAD2* genes and concluded that the changes in fatty acid composition were a combination of somatic cell and germline cell mutations. The French group[13] were able to obtain a wide assortment of *FAD2* gene knockouts that lead to a variety of fatty acid profile phenotypes, including oleic acid concentrations above 60%. Indeed, they were able to obtain plants containing homozygous knockouts of all six *FAD2* genes. However, such plants displayed poor growth characteristics and pointed to the importance of saturated fatty acids for the normal growth and development of Camelina. The work from both groups provided proof of concept that the *FAD2* genes in Camelina can be successfully edited using the CRISPR/Cas9 system to yield plants capable of producing commercially valuable oils and that gene editing to obtain other valuable traits in this increasingly important allohexaploid oil seed crop should be possible.

2.2.2 Creating an Herbicide Resistance Trait in Oil Seed Rape and Canola Using ssODNs

The ability to change nucleotide sequences within target genes is a powerful means of creating important new genetic traits in plants and other organisms. Although developed for use in plants more than 15 years ago,[34–37] the use of single-stranded desoxyoligonucleotides (ssODNs) with protected 5′ and 3′ ends has been shown only recently to be of practical importance for creating useful traits in crop plants (e.g., Refs. 37a,b). This capability was shown to extend to polyploid crops when Gocal *et al.*[14] demonstrated that ssODNs could be used to make specific nucleotide changes in the oil seed rape acetolactate synthase (*ALS*) gene to cause production of an acetolatate synthase that was not inhibited by the sulfonylurea class of herbicides. For this research, they used a 5′Cy3 label and a 3′idC reverse base-protected ssODN design that allowed detection of the oligonucleotide within cells by virtue of the Cy3 label. Three ssODNs were constructed that, respectively, targeted the P197, W574, and S653 amino acid residues of ALS for modification for amino acid substitutions that were known to block sensitivity of the ALS enzyme to sulfonyl urea herbicides. This herbicide resistance trait has been incorporated into canola and is presently being marketed by Cibus, Inc., in North America. Importantly, the ssODN used to stimulate mutations is short lived in the cell and the resulting crop plant contains no foreign DNA—and, thus, has been exempted from regulation by the USDA because it is not classified as a GMO.

2.3 Potato

2.3.1 Potato Is Readily Amenable to Gene Editing With TALENs and CRISPR/Cas9

Potatoes (*Solanum tuberosum*) are the world's fourth largest food crop and have been dispersed from their origin in South American to many countries around the world. Although introgression of traits such as disease resistance and pest resistance from wild varieties of potatoes is possible, the tetraploid nature of potatoes makes this a challenging and time-consuming task. Thus, an important breakthrough in 2015 was the demonstration that both TALENs[38] and CRISPR/Cas9[39] gene-editing systems could be used to efficiently modify the potato genome. It was further demonstrated by Forsyth *et al.*[15] that the TALEN system could be used to successfully target T-DNA incorporation into a specific prechosen site in the potato genome that is transcriptionally active. Importantly, these investigators designed a vector that would not allow stable integration of the TALEN genes into the genome. Their data indicated that TALEN-induced T-DNA integration can occur with high frequency and results in consistent expression of the integrated gene of interest.

2.3.2 Use of TALEN Genes and Donor DNA Fragments Delivered by Geminivirus Replicons to Stimulate HR and Creation of Herbicide Tolerant Potatoes

As described earlier for use in wheat,[10] geminivirus-based DNA replicons can also be used to make genome edits that would be difficult using standard gene-editing approaches. In particular, Butler *et al.*[16] have demonstrated that a single-stranded geminivirus-based DNA replicon can be created that carries TALEN genes targeting a specific site in the *ALS* gene of potato and, as well, a fragment of the *ALS* gene that carries a mutation known from previous work to confer tolerance to several classes of ALS-inhibiting herbicides (e.g., sulfonylureas). When potato cells are transfected with the geminivirus DNA replicon construct, the TALEN genes were transiently expressed and caused a DSB at the target site within the *ALS* gene. The presence of a DSB allowed the cell's system for DNA repair by HR to recognize the ssDNA fragment of the *ALS* gene carrying the desired mutation and integrate this new sequence in place of the wild-type *ALS* sequence. Plants regenerated from such transient transformations (i.e., no TALEN or geminivirus DNA sequences persisted in the regenerated plants) were shown to be significantly more tolerant to herbicide treatments. Gene or gene fragment replacement by HR is a powerful technique that should allow significant changes to be incorporated into plant genomes—an especially valuable tool for crops like potato that are primarily vegetatively propagated.

2.3.3 Improving Cold Storage and Processing Traits in Potato Through Targeted Gene Knockout

Cold storage of potatoes is important for extending the postharvest shelf life of potatoes. However, such storage causes increases in reducing sugars that can, upon frying of the stored potato, lead to reactions with amino acids. Such reactions can promote undesirable browning, creation of bitter tastes, and production of low amounts of toxic acrylamide. To avoid conversion of sucrose to the reducing sugars, glucose and fructose, Clasen *et al.*[17] targeted the vacuolar invertase (*Vinv*) genes of Ranger Russet potatoes for knockout using the TALEN gene-editing system. Five out of 18 regenerated plants contained knockouts of all four *Vinv* alleles. Tubers from these plants contained no detectable reducing sugars and were light brown after frying. Importantly, they also contained lower levels of acrylamide and 7 of the 18 regenerated plants were shown to contain no TALEN genes.

2.3.4 Modification of Starch Composition of Potatoes Using TALENs or CRISPR/Cas9 to Edit the Four Alleles of the Granule-Bound Starch Synthase Gene

Potato starch provides important nutrition for humans and animals as well as being utilized for coatings in numerous industrial processes. The relative balance of the two major starch types, amylose and amylopectin, dictates which type of potato starch is best for each commercial application. Thus, the ability to carefully control this balance may have significant value. As an initial demonstration of the ability to edit one key enzyme in the starch biosynthesis pathway, granule-bound starch synthase (GBSS), Kusano *et al.*[18] used the TALEN system to successfully disrupt copies of this gene in potato protoplast cells. In a more recent and extensive study, Andersson *et al.*[19] used transient expression of the CRISPR/Cas9 system to demonstrate complete knockout of all four GBSS alleles in PEG-treated potato protoplasts and in up to 2% of regenerated lines. Complete knockout of the GBSS genes resulted in the complete lack of amylose in regenerated potato microtubers.

2.3.5 Potential for New Approaches to Disease Resistance in Potatoes

Although not yet accomplished in potatoes, there is great anticipation that gene-editing techniques can soon be used to devise new and perhaps novel approaches to creating resistance to plant viruses, bacteria, and fungi (e.g., Refs. 20,40,41). Of special interest[20] is *Phytophthora infestans*, the cause of potato blight or late blight, where it is hoped that use of gene-editing techniques can potentially be used for creation of new disease resistances by knocking out or removing disease susceptibility genes (S-genes).

2.4 Cotton

2.4.1 CRISPR/Cas9 as a System to Help Control Cotton Leaf Curl Disease and Other Viral Infections

The use of the CRISPR/Cas9 gene-editing system has been proposed as a tool to help control cotton leaf curl disease, in particular,[21] and plant viruses, in general.[41] The begomoviruses (family Geminiviridae) associated with cotton leaf curl disease (CLCuD) are a major threat to cotton production in India, Pakistan, and other parts of Asia. Methods to control these viruses depend not only on controlling production of the circular, ssDNA genome but also two satellite DNAs, alphasatellite and betasatellite. Given success in using the CRISPR/Cas9 system to control other geminiviruses, it has been proposed that a unique, broad-spectrum multiplexed CRISPR/Cas9 system can be produced in which sgRNA are designed to target not only the whole CLCuD-associated begomovirus complex but also its satellite DNAs.

2.5 Peanut

2.5.1 Important Target Genes in Peanut—As Identified by RNA Interference

Peanuts are an important source of protein and fatty acids in the diets of people in many parts of the world. However, peanut plants are susceptible to numerous plant diseases, fungal infections, and soil nutrition challenges, including maintenance of favorable symbiotic relations with nitrogen-fixing bacteria. Peanuts, themselves, suffer from containing storage proteins that are major food allergens for a small, but significant, portion of consumers. No reports of use of gene-editing techniques in peanut have yet appeared. However, there appear no important technical reasons why such approaches cannot be applied to peanuts. Indeed, a number of genes that would appear to be prime targets for knockout or editing can already be nominated based on the success of several investigators in using RNA interference to partially or completely silence expression of important genes. Examples include reduction of peanut allergen production in peanut seeds,[22,42,42a] reduction in growth of *Aspergillus flavus*, a major producer of aflatoxin,[23,42a] and identification of key components of nodule/nitrogen-fixing bacteria symbiosis (e.g., Ref. 43).

2.6 Sugar Cane

2.6.1 TALEN-Mediated Reduction in Lignin Content in Sugarcane to Improve Efficiency of Biofuel Production

Sugar cane is one of the world largest sources of food-grade sucrose and an increasingly important source of sugar for ethanol biofuel production. As such, it would appear to be a prime target for genetic engineering, including gene editing. However, sugar cane's quite large genome, its propensity to silence transgenes and its high polyploidy ($x=10$–13) and functional redundancy among homoeologs are important obstacles to overcome. In spite of these challenges, Jung and Altpeter[24] were able to successfully target the multiple caffeic acid *O*-methyltransferase (COMT) genes using TALENs directed against a highly conserved region of this gene. The level of reduction in lignin content among regenerated sugar cane lines was positively correlated with the frequency of COMT gene knockout in each plant. Indeed, plants showing the highest levels of gene knockout (~99%) contained 29%–32% lower levels of lignin than progenitor plants.

2.7 Citrus

2.7.1 Knockout of the Citrus CsLOB1 Gene Using CRISPR/Cas9 Confers Resistance to Citrus Canker

Citrus production around the world is susceptible to challenges including diseases such as bacterial canker and Huanglongbing (HLB). Breeding programs can address some of these challenges but suffer from long crossing cycles, extended juvenility, polyembyrony and last, but not the least, polyploidy. Use of gene-editing techniques hold promise that the obstacles created by polyploidy can be overcome and allow scientists to achieve pivotal changes in target genes in a relatively short period of time. Evidence that this can be the case is provided by the recent work of Jia *et al.*[25] who used CRISPR/Cas9 technology to target the canker susceptibility gene, *CsLOB1*, of Duncan grapefruit for knockout. They achieved high rates of gene modification in most of the six independent knockout lines they created. Two of these lines showed significantly reduced infection with the citrus canker-causing pathogen *Xanthomonas citri* subsp. citri (Xcc) and the small pustules that developed on these plants did not develop into typical canker symptoms. Thus, the CRISPR/Cas9 system appears to hold great promise to help control and contain the spread of devastating crop diseases such as citrus canker.

2.8 Apple and Other Fruit

Breeding of polyploid fruit trees is among the most difficult and time-consuming challenges to fruit crop improvement—but is important to development of new varieties with higher quality, enhanced nutrition, and improved agronomic traits such as disease resistance and drought tolerance. As with the other crops discussed earlier, gene-editing techniques are seen as a potentially valuable means of speeding the introduction of new traits into existing fruits and fruit trees. An early attempt to demonstrate that gene editing was possible in apple was by Peer *et al.*[44] who used ZFNs to target a mutated transgene, *uidA*, to cause nucleotide indels that, in some cases, would restore the proper reading frame to the *uidA* gene and allow production of a purple (GUS) stain in cells containing the successfully reverted *uidA* gene. This was followed by work of Nishitani *et al.*[45] who used the CRISPR/Cas9 editing technique to successfully target the endogenous phytoene desaturase (*PDS*) gene of apple. This group was able to demonstrate production of regenerated plantlets with albino leaf areas that are expected when PDS is not active and chlorophyll synthesis is strongly diminished. Mutations in the *PDS* genes were confirmed by DNA sequencing.

An important advance was made by Malnoy *et al.*[26] who recently demonstrated that gene editing in golden delicious apple protoplasts could be achieved using the same kinds of Cas9/sgRNA RNP complexes discussed earlier in reference to gene editing in other plants.[11,32] This group was able to successfully target the *DIPM-1, DIPM-2*, and *DIPM-4* genes in apple protoplasts to potentially increase resistance to fire blight disease in regenerated apple plants. From this work, we can see that delivery of Cas9/sgRNA RNP complexes can efficiently support gene editing in fruit trees leaving only the desired mutant gene and no foreign DNA. This provides a degree of encouragement that the burdensome regulatory system that mires traditional GMOs, with luck, may be avoided or minimized for minor or specialty crops that cannot support the present time-consuming and money-consuming regulatory processes.

REFERENCES

1. Jiao R, Gao C. The CRISPR/Cas9 genome editing revolution. *J Genet Genomics*. 2016;43(5):227–228. http://dx.doi.org/10.1016/j.jgg.2016.05.004.
2. Puchta H. Applying CRISPR/Cas for genome engineering in plants: the best is yet to come. *Curr Opin Plant Biol*. 2017;36:1–8. http://dx.doi.org/10.1016/j.pbi.2016.11.011.
3. Weeks DP, Spalding MH, Yang B. Use of designer nucleases for targeted gene and genome editing in plants. *Plant Biotechnol J*. 2016;14(2):483–495. http://dx.doi.org/10.1111/pbi.12448.
4. Comai L. The advantages and disadvantages of being polyploid. *Nat Rev Genet*. 2005;6(11):836–846. http://dx.doi.org/10.1038/nrg1711.
5. Harper AL, Trick M, Higgins J, et al. Associative transcriptomics of traits in the polyploid crop species Brassica napus. *Nat Biotechnol*. 2012;30(8):798–802. http://dx.doi.org/10.1038/nbt.2302.
6. Mason AS, Snowdon RJ. Oilseed rape: learning about ancient and recent polyploid evolution from a recent crop species. *Plant Biol (Stuttg)*. 2016;18(6):883–892. http://dx.doi.org/10.1111/plb.12462.
7. Pont C, Salse J. Wheat paleohistory created asymmetrical genomic evolution. *Curr Opin Plant Biol*. 2017;36:29–37. http://dx.doi.org/10.1016/j.pbi.2017.01.001.
7a. Upadhyay SK, Kumar J, Alok A, Tuli R. RNA-guided genome editing for target gene mutations in wheat. *G3 (Bethesda)*. 2013;3(12):2233–2238. https://dx.doi.org/10.1534/g3.113.008847.
8. Wang Y, Cheng X, Shan Q, et al. Simultaneous editing of three homoeoalleles in hexaploid bread wheat confers heritable resistance to powdery mildew. *Nat Biotechnol*. 2014;32(9):947–951. http://dx.doi.org/10.1038/nbt.2969.

9. Zhang Y, Liang Z, Zong Y, et al. Efficient and transgene-free genome editing in wheat through transient expression of CRISPR/Cas9 DNA or RNA. *Nat Commun.* 2016;7:12617. http://dx.doi.org/10.1038/ncomms12617.
10. Gil-Humanes J, Wang Y, Liang Z, et al. High efficiency gene targeting in hexaploid wheat using DNA replicons and CRISPR/Cas9. *Plant J.* 2016;89(6):1251–1262. http://dx.doi.org/10.1111/tpj.13446..
11. Liang Z, Chen K, Li T, et al. Efficient DNA-free genome editing of bread wheat using CRISPR/Cas9 ribonucleoprotein complexes. *Nat Commun.* 2017;8:14261. http://dx.doi.org/10.1038/ncomms14261.
12. Jiang WZ, Henry IM, Lynagh PG, Comai L, Cahoon EB, Weeks DP. Significant enhancement of fatty acid composition in seeds of the allohexaploid, Camelina sativa, using CRISPR/Cas9 gene editing. *Plant Biotechnol J.* 2017;15:648–657. http://dx.doi.org/10.1111/pbi.12663.
13. Morineau C, Bellec Y, Tellier F, et al. Selective gene dosage by CRISPR-Cas9 genome editing in hexaploid Camelina sativa. *Plant Biotechnol J.* 2017;15(6):729–739. http://dx.doi.org/10.1111/pbi.12671.
14. Gocal GFW, Schopke C, Beetham PR. Oligo-mediated targeted gene editing. In: Zhang F, Puchta H, Thomson JG, eds. *Advances in New Technology for Targeted Modification of Plant Genomes.* Berlin, Heidelberg: Springer Verlag; 2015:73–90. Chapter 5.
15. Forsyth A, Weeks T, Richael C, Duan H. Transcription activator-like effector nucleases (TALEN)-mediated targeted DNA insertion in potato plants. *Front Plant Sci.* 2016;7:1572. eCollection 2016.
16. Butler NM, Baltes NJ, Voytas DF, Douches DS. Geminivirus-mediated genome editing in potato (Solanum tuberosum L.) using sequence-specific nucleases. *Front Plant Sci.* 2016;7:1045. http://dx.doi.org/10.3389/fpls.2016.01045.
17. Clasen BM, Stoddard TJ, Luo S, et al. Improving cold storage and processing traits in potato through targeted gene knockout. *Plant Biotechnol J.* 2016;14(1):169–176. http://dx.doi.org/10.1111/pbi.12370.
18. Kusano H, Onodera H, Kihira M, Aoki H, Matsuzaki H, Shimada HA. Simple gateway-assisted construction system of TALEN genes for plant genome editing. *Sci Rep.* 2016;6:30234. http://dx.doi.org/10.1038/srep30234.
19. Andersson M, Turesson H, Nicolia A, Fält AS, Samuelsson M, Hofvander P. Efficient targeted multiallelic mutagenesis in tetraploid potato (Solanum tuberosum) by transient CRISPR-Cas9 expression in protoplasts. *Plant Cell Rep.* 2017;36(1):117–128. http://dx.doi.org/10.1007/s00299-016-2062-3.
20. Schaart JG, van de Wiel CC, Lotz LA, Smulders MJ. Opportunities for products of new plant breeding techniques. *Trends Plant Sci.* 2016;21(5):438–449. http://dx.doi.org/10.1016/j.tplants.2015.11.006.
21. Iqbal Z, Sattar MN, Shafiq M. CRISPR/Cas9: a tool to circumscribe cotton leaf curl disease. *Front Plant Sci.* 2016;7:475. http://dx.doi.org/10.3389/fpls.2016.00475.
22. Chandran M, Chu Y, Maleki SJ, Ozias-Akins P. Stability of transgene expression in reduced allergen peanut (Arachis hypogaea L.) across multiple generations and at different soil sulfur levels. *J Agric Food Chem.* 2015;63(6):1788–1797. http://dx.doi.org/10.1021/jf504892f.
23. Arias RS, Dang PM, Sobolev VS. RNAi-mediated control of aflatoxins in peanut: method to analyze mycotoxin production and transgene expression in the peanut/aspergillus pathosystem. *J Vis Exp.* 2015;106:e53398. http://dx.doi.org/10.3791/53398.
24. Jung JH, Altpeter F. TALEN mediated targeted mutagenesis of the caffeic acid O-methyltransferase in highly polyploid sugarcane improves cell wall composition for production of bioethanol. *Plant Mol Biol.* 2016;92(1–2):131–142. http://dx.doi.org/10.1007/s11103-016-0499-y.

25. Jia H, Zhang Y, Orbović V, et al. Genome editing of the disease susceptibility gene CsLOB1 in citrus confers resistance to citrus canker. *Plant Biotechnol J.* 2016;14: 1291–1301. http://dx.doi.org/10.1111/pbi.12677.
26. Malnoy M, Viola R, Jung MH, et al. DNA-free genetically edited grapevine and apple protoplast using CRISPR/Cas9 ribonucleoproteins. *Front Plant Sci.* 2016;7:1904. http:// dx.doi.org/10.3389/fpls.2016.01904.
26a. Li T, Liu B, Spalding MH, Weeks DP, Yang B. High-efficiency TALEN-based gene editing produces disease-resistant rice. *Nat Biotechnol.* 2012;30(5):390–392. http://dx. doi.org/10.1038/nbt.2199.
27. Waltz EA. Face-lift for biotech rules begins. *Nat Biotechnol.* 2015;33(12):1221–1222. http://dx.doi.org/10.1038/nbt1215-1221.
28. Waltz E. CRISPR-edited crops free to enter market, skip regulation. *Nat Biotechnol.* 2016;34(6):582. http://dx.doi.org/10.1038/nbt0616-582.
29. Waltz E. Gene-edited CRISPR mushroom escapes US regulation. *Nature.* 2016;532(7599):293. http://dx.doi.org/10.1038/nature.2016.19754.
30. Baltes NJ, Gil-Humanes J, Cermak T, Atkins PA, Voytas DF. DNA replicons for plant genome engineering. *Plant Cell.* 2014;26(1):151–163. http://dx.doi.org/10.1105/tpc. 113.119792.
31. Kim S, Kim D, Cho SW, Kim J, Kim JS. Highly efficient RNA-guided genome editing in human cells via delivery of purified Cas9 ribonucleoproteins. *Genome Res.* 2014;24:1012–1019. http://dx.doi.org/10.1101/gr.171322.113.
32. Woo JW, Kim J, Kwon SI, et al. DNA-free genome editing in plants with preassembled CRISPR-Cas9 ribonucleoproteins. *Nat Biotechnol.* 2015;33:1162–1164. http://dx.doi. org/10.1038/nbt.3389.
33. Pacher M, Puchta H. From classical mutagenesis to nuclease-based breeding—directing natural DNA repair for a natural end-product. *Plant J.* 2017;90:819–833. http://dx.doi. org/10.1111/tpj.13469.
33a. Svitashev S, Schwartz C, Lenderts B, Young JK, Cigan AM. Genome editing in maize directed by CRISPR-Cas9 ribonucleoprotein complexes. *Nat Commun.* 2016;7: 13274–13282. http://dx.doi.org/10.1038/ncomms13274.
34. Beetham PR, Kipp PB, Sawycky XL, Arntzen CJ, May GD. A tool for functional plant genomics: chimeric RNA/DNA oligonucleotides cause in vivo gene-specific mutations. *Proc Natl Acad Sci USA.* 1999;96(15):8774–8778.
35. Kochevenko A, Willmitzer L. Chimeric RNA/DNA oligonucleotide-based site-specific modification of the tobacco acetolactate syntase gene. *Plant Physiol.* 2003;132(1): 174–184.
36. Ruiter R, van den Brande I, Stals E, Delauré S, Cornelissen M, D'Halluin K. Spontaneous mutation frequency in plants obscures the effect of chimeraplasty. *Plant Mol Biol.* 2003;53(5):675–689.
37. Zhu T, Peterson DJ, Tagliani L, St Clair G, Baszczynski CL, Bowen B. Targeted manipulation of maize genes in vivo using chimeric RNA/DNA oligonucleotides. *Proc Natl Acad Sci USA.* 1999;96(15):8768–8773.
37a. Sauer NJ, Narváez-Vásquez J, Mozoruk J, et al. Oligonucleotide-mediated genome editing provides precision and function to engineered nucleases and antibiotics in plants. *Plant Physiol.* 2016;170(4):1917–1928. http://dx.doi.org/10.1104/pp.15.01696.
37b. Sauer NJ, Mozoruk J , Miller RB, et al. Oligonucleotide-directed mutagenesis for precision gene editing. *Plant Biotechnol J.* 2016;14(2):496–502. http://dx.doi.org/10.1111/ pbi.12496.
38. Nicolia A, Proux-Wéra E, Åhman I, et al. Targeted gene mutation in tetraploid potato through transient TALEN expression in protoplasts. *J Biotechnol.* 2015;204:17–24. http://dx.doi.org/10.1016/j.jbiotec.2015.03.021.

39. Wang S, Zhang S, Wang W, Xiong X, Meng F, Cui X. Efficient targeted mutagenesis in potato by the CRISPR/Cas9 system. *Plant Cell Rep*. 2015;34(9):1473–1476. http://dx.doi.org/10.1007/s00299-015-1816-7.
40. Sundin GW, Castiblanco LF, Yuan X, Zeng Q, Yang CH. Bacterial disease management: challenges, experience, innovation and future prospects: challenges in bacterial molecular plant pathology. *Mol Plant Pathol*. 2016;17(9):1506–1518. http://dx.doi.org/10.1111/mpp.12436.
41. Zaidi SS, Tashkandi M, Mansoor S, Mahfouz MM. Engineering plant immunity: using CRISPR/Cas9 to generate virus resistance. *Front Plant Sci*. 2016;7:1673. eCollection 2016.
42. Dodo HW, Konan KN, Chen FC, Egnin M, Viquez OM. Alleviating peanut allergy using genetic engineering: the silencing of the immunodominant allergen Ara h 2 leads to its significant reduction and a decrease in peanut allergenicity. *Plant Biotechnol J*. 2008;6(2):135–145.

42a. Chu Y, Faustinelli P, Ramos ML, et al. Reduction of IgE binding and nonpromotion of Aspergillus flavus fungal growth by simultaneously silencing Ara h 2 and Ara h 6 in peanut. *J Agric Food Chem*. 2008;56(23):11225–11233. http://dx.doi.org/10.1021/jf802600r.

43. Sinharoy S, DasGupta MRNA. Interference highlights the role of CCaMK in dissemination of endosymbionts in the Aeschynomeneae legume Arachis. *Mol Plant Microbe Interact*. 2009;22(11):1466–1475. http://dx.doi.org/10.1094/MPMI-22-11-1466.
44. Peer R, Rivlin G, Golobovitch S, et al. Targeted mutagenesis using zinc-finger nucleases in perennial fruit trees. *Planta*. 2015;241(4):941–951. http://dx.doi.org/10.1007/s00425-014-2224-x. Epub 2014 Dec 21.
45. Nishitani C, Hirai N, Komori S, et al. Efficient genome editing in apple using a CRISPR/Cas9 system. *Sci Rep*. 2016;6:31481. http://dx.doi.org/10.1038/srep31481.

CHAPTER FIVE

Gene Editing With TALEN and CRISPR/Cas in Rice

Honghao Bi, Bing Yang[1]
Iowa State University, Ames, IA, United States
[1]Corresponding author: e-mail address: byang@iastate.edu

Contents

Abstract

Engineered, site-specific nucleases induce genomic double-strand DNA breaks and break repair processes enable genome editing in a plethora of eukaryotic genomes. TALENs (transcription activator-like effector nucleases) and CRISPR/Cas (clustered regularly interspaced short palindromic repeats and CRISPR-associated proteins) are potent biotechnological tools used for genome editing. In rice, species-tailored editing tools have proven to be efficient and easy to use. Both tools are capable of generating DNA double-strand breaks (DSBs) in vivo and such breaks can be repaired either by error-prone NHEJ (nonhomologous end joining) that leads to nucleotide insertions or deletions or by HDR (homology-directed repair) if an appropriate exogenous DNA template is provided. NHEJ repair often results in gene knockout, while HDR results in precise nucleotide sequence or gene replacement. In this review, we revisit the molecular mechanisms underlying DSB repair in eukaryotes and review the TALEN and CRISPR technologies (CRISPR/Cas9, CRISPR/Cpf1, and Base Editor) developed and utilized for genome editing by scientists in rice community.

Progress in Molecular Biology and Translational Science, Volume 149
ISSN 1877-1173
http://dx.doi.org/10.1016/bs.pmbts.2017.04.006

1. INTRODUCTION

Rice (*Oryza sativa*), a crop that provides staple food for more than half of the world's population, has served as an excellent model plant for functional genomic studies. This is due to its richness of genetic resources, genomic synteny with other cereals, amenable transformability, and a rich history of extensive research efforts. Rice has also provided a model for establishing technological platforms and testing strategies for genetic improvements to itself and to other crops.[1,2] For the past decades, we have witnessed tremendous progress in gene discovery and genomic analysis of agronomically important traits and biological processes in rice, largely thanks to the availability of genetic resources and tools.[3,4] The complex genetic makeup of rice, as with other crop species, allows continued progress in increasing yields, fending off biotic and abiotic stresses and improving nutritional quality—the ultimate goals in breeding elite rice. This genetic diversity allows novel and improved genetic traits fueled by new discoveries and synthetic biology. Biotechnologies, especially those related to altering the genetic makeup of rice varieties, have played and will continue to play an important role not only in the basic understanding of rice biology but also in the creation and introgression of novel genetic elements for improved agronomic traits.[3] Knowledge obtained from rice as a model system can also be extrapolated to other cereal crops, in particular, and to all other crop plants, in general.

Recent advances in genome-editing technology allow precise, targeted genomic changes, including whole-gene insertion or deletion, stacking or pyramiding of genes, and precise modification of genetic elements that allow scientists to introduce, change, or optimize genetic traits of interest. Several classes of edit-enabling technologies have been developed in the past decades, including zinc finger nucleases, TAL effector nucleases (TALENs) and most recently clustered regularly interspaced short palindromic repeats (CRISPR)/CRISPR-associated proteins (CRISPR/Cas).[5] The diverse genetic toolboxes, especially TALEN and CRISPR, represent the state-of-the-art genome-editing technologies and hold immense promise for defining the genetic and molecular basis for valuable traits, as well as for the introgression of valuable genes/traits into cultivated varieties in a timely and economically sound manner. The aims of this review chapter are, therefore, to provide a clear understanding of the molecular mechanisms involved in TALEN and CRISPR/Cas9 gene editing and to highlight the research endeavors in the plant science community for their utilization in rice improvement.

2. CREATION OF DNA DOUBLE-STRAND BREAKS TO ALLOW GENOME EDITING

Preselected and site-specific DNA double-strand breaks (DSBs) are a prerequisite for effective genome editing in eukaryotes. Present methods for creating DSBs utilize engineered nucleases and rely on subsequent DNA repair mechanisms that are shared by all eukaryotic organisms. Repair to DSBs through error-prone nonhomologous end joining (NHEJ) DNA repair is the predominant mechanism utilized in cells and is the primary means for creating gene knockouts. Homology-directed repair (HDR)-mediated gene replacement or insertion is intrinsically less efficient and is technically more challenging—but is of potentially great significance for crop improvement. Indeed, a better understanding of the molecular mechanisms underlying DSB repair and the HDR process are important goals for designing and achieving more efficient gene replacement in plants.[6]

DNA damage such as DSBs triggers a series of cellular processes including DNA damage response and cell cycle checkpoints, eventually leading to DNA damage repair.[7] Repair to DSBs through NHEJ involves the binding of DSB ends by the DNA-bound Ku proteins, recruiting and forming a DNA–protein kinase complex, end processing, and eventually NHEJ facilitated by multiple enzyme complexes such as the DNA ligation complex.[8,9] NHEJ occurs at any stage of cell development and during any phase of cell division in higher eukaryotes and leads to low fidelity DNA repair.[10] Such repair often leads to limited or extensive addition or deletion of nucleotides and, thus, DNA sequence alterations at the rejoined sites—a desired outcome when one wishes to create gene knockouts.

In contrast, HDR can occur at a DSB site in the presence of homologous DNA template, often leading to a faithful damage repair. The template for HDR could be a DNA fragment on a homologous chromosome or on the sister chromatid generated after DNA replication in higher eukaryotes. As such, the latter type of HDR occurs mainly during the S and G2 phases of the cell cycle when fully formed sister chromatids are available.[10] In the repair process, each 5′ end of the DSB is resected, resulting in a 3′ extended region homologous to the donor DNA template and subsequent D-loop formation between both. With the free 3′ end of the DSB as a primer and the matching strand of donor DNA as template, DNA polymerases extend the length of resected DNA strand followed by the ligation of the DNA ends and generation of a Holliday junction and a heteroduplex molecule. Finally, resolution of Holliday junctions results in either crossover or noncrossover products.[11] Importantly for crop improvement purposes, the homologous DNA template can be an exogenous DNA strand containing a desired genetic change that can be supplied along with the DSB-causing reagents, such as TALENs or CRISPR/Cas9, for the purpose of gene replacement.

3. USE OF TALENs FOR GENE EDITING IN RICE

TAL effectors (TALEs) from bacteria of the *Xanthomonas* genus represent the largest type III effector proteins and play an important role in bacterial pathogenesis of host plants.[12,13] TALEs, once internalized into the host cell milieu, function to transcriptionally activate host gene expression by binding to the promoters of target genes in a sequence-specific manner.[14,15] The central repeats of usually 34 amino acids form a supersolenoid wrapping around the double helix of target DNA with the two amino acids at the positions 12 and 13 determining the specificities of each repeat corresponding to a particular nucleotide in the target DNA sequence.[16,17] The so-called DNA recognition code is manifested as one repeat unit of TALE binding to one nucleotide of target DNA with each type of the various repeats preferentially binding to one of the four different types of nucleotide (A, G, C, or T). The discovery of the so-called DNA recognition code made possible the engineering DNA-binding proteins based on preselected DNA sequences by using the four prominent repeats each recognizing to one of the four different nucleotides.[18,19] Thus, assembly of this set of four different repeat elements could be designed to allow precise matching to any given DNA sequence from any given organism. As a result, such custom engineered DNA-binding proteins enabled creation of biotechnological tools for genome editing and gene regulation.[20] The most predominant utilization of TAL DNA-binding domains is to engineer programmable TALENs for induction of DSBs that are exploitable for site-specific genome editing in eukaryotes.[21,22] TALENs are produced by fusing TALE DNA-binding domains (either full-length TALEs or only central repeats) and the *Fok*I nuclease cleavage domain.[23,24] TALENs function in pairs to make DSBs as one binds to one DNA strand and the other binds to the adjacent complementary strand with a spacer of about 20 bp wherein the dimerizing *Fok*I domains introduce a DSB. Each TALEN usually defines a specificity of 16–24 bp sequence dependent on the preference of users, and the paired TALENs achieve a combined specificity of 32–48 bp at the target site (Fig. 1A).

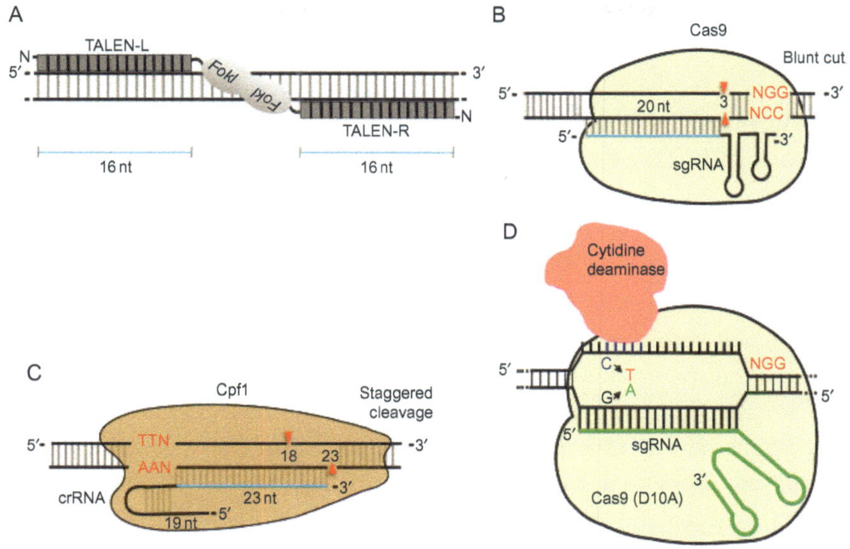

Fig. 1 Engineered nucleases predominantly used for rice genome editing. Nucleases form a complex with the target genomic dsDNA to carry out their catalytic activity. (A) Paired TALENs bind to the opposite strands in a way to enable the dimerizing *FokI* DNA cleavage domains to cause DSBs. (B) sgRNA guides Cas9 to the user-chosen site to induce a blunt DSB 3-bp upstream of the PAM sequence. (C) crRNA guides Cpf1 to the genomic locus to introduce a staggered cleavage 18–23 nt downstream of the 5′ located PAM sequence. (D) crRNA-guided Cas9–cytidine deaminase converts cytidine to uracil, resulting in the C–G conversion to T–A within the window of five nucleotides 13 bp upstream of the PAM site at the targeted genomic site.

The first step of TALEN utilization is to custom engineer a TAL repeat array. Various repeat assembly methods have been developed based on the modular nature of four repeats.[18,25–28] The majority of TALEN applications in rice, and other plant species as well, has focused on targeted mutagenesis.[22] The first TALEN-modified rice was made by mutating the *OsSWEET14* promoter to generate heritable, disease-resistant lines.[29] In that work,[29] and as a proof-of-concept application with TALEN technology, two pairs of TALENs, including one TALEN that was derived from naturally occurring TALE—AvrXa7, were engineered to target the promoter element in the sucrose efflux transporter *OsSWEET14* gene. The so-called effector binding element (EBE) bound by the natural TALEs (AvrXa7 and PthXo3) from *Xanthomonas oryzae* pv. *oryzae* (Xoo), causal pathogen of bacterial blight of rice.[30] High efficiencies of mutagenesis by the two pairs of TALENs were achieved in T0 generation plants; inheritability of mutations and TALEN transgene segregation were also demonstrated in the progeny plants. The mutant plants homozygous for the EBE mutations were highly resistant to Xoo strains that depended on AvrXa7 to induce *OsSWEET14*-associated bacterial blight.[29] Similarly, the promoter EBEs in *OsSWEET14* targeted by TalC and Tal5, two TALEs from the African Xoo strains, were subjected to targeted mutagenesis with engineered TALENs. The resulting mutations in the respective EBEs at high frequencies provided strong blight resistance.[31]

TALENs were also used by several other groups for targeted mutagenesis in rice. Examples include mutagenizing lipoxygenase *LOX3*, resulting in seed storage improvement[32]; knockout of *OsBADH2* (betaine aldehyde dehydrogenase 2), leading to increased amounts of the desirable fragrant compound, 2-acetyl-1-pyrroline in grain.[33] Additionally, TALENs were used to validate and expand prior studies on DSB-induced small RNA (diRNA) biogenesis and the role of these molecules in DSB repair.[34] In this study, engineered TALENs were used to introduce DSBs in transgenes and endogenous genes, resulting in findings that DSB and highly transcribed mRNAs are needed for diRNA production and, importantly, that abundant diRNAs are not essential for DSB repair.[34] Finally, TALENs have been demonstrated to facilitate the HDR-mediated gene editing or gene replacement in rice.[35]

4. USE OF CRISPR/CAS9 FOR GENE EDITING IN RICE

CRISPR/Cas9 is a ribonucleoprotein complex consisting of a single effector protein, Cas9, and a CRISPR RNA. CRISPR/Cas9 is an adaptive immune system that is widespread in bacteria and archaea and is used to protect hosts from infection by invading viruses, phages, and plasmids.[36] The CRISPR locus was first identified and reported as an array of 29-bp tandem repeats interspaced with 32 bp spacer sequences of DNA in *Escherichia coli* in 1987.[37] Later, Mojica *et al.* reported their discovery of structurally similar direct repeat sequences of 24–40 bp in various physiologically and phylogenetically distant bacteria and archaea. They found that prokaryotes from the same phylogenetic group contained highly conserved repeats, but unique 20–58 bp long spacer sequences.[38,39] In 2002, Jansen *et al.* performed in silico analysis on the sequenced genomes of archaea and bacteria and defined this type of repetitive sequences as CRISPR; they also identified the CRISPR-associated (cas) genes in close proximity to the CRISPR sequences.[40] It was not until 2006 that the biological significance of CRISPR/Cas system was hypothesized to play a role in the immunity of prokaryotes to virus infection.[41] This hypothesis was later validated by experimentation.[42–44]

The most significant breakthrough in CRISPR research came in 2012 with the work by Jinek *et al.* demonstrating in vitro DNA cleavage activity using recombinant Cas9 coupled with a single guide RNA (sgRNA) derived from the fusion of CRISPR RNA (crRNA) and *trans*-activating crRNA (tracrRNA).[45] The simplified CRISPR/Cas9 system based on the *Streptococcus pyrogenes* CRISPR locus consists of one effector component (SpCas9) and a sgRNA that contains a Cas9 recognizable scaffold (~85 nucleotides) preceded with a 5′ end 20-nt sequence complementary to the target DNA sequence[45] (Fig. 1B). The potential of CRISPR/Cas9 for gene editing was later substantiated by two seminal works using human cultured cells.[46,47] Several laboratories later in 2013 then demonstrated the feasibility of CRISPR/Cas9-mediated gene editing in plant cells and production of stable transgenic plants containing successfully edited genes.[48–55] From that time forward, the number of publications describing the development and utilization of CRISPR/Cas9, its variants, and alternative systems for genome editing in plants has increased dramatically.

Rice, as no surprise, was among the first plants to be used to demonstrate the feasibility of CRISPR-mediated targeted mutagenesis[48,49,51,53,56] and gene replacement.[57,58] It also was used to optimize components of CRISPR/Cas9,[59] and develop alternative CRISPR/Cas9 platforms (e.g., SpCas9 mutants containing relaxed PAM sequence requirements, SpCas9 nickase, and SaCas9).[60–62] In these early uses of CRISPR/Cas9 for targeted mutagenesis, the delivery of Cas9 and guide RNA expression cassettes was achieved using *Agrobacterium tumefaciens* cells containing a single binary vector, while for gene replacement experiments template DNA and plasmid DNA carrying CRISPR/Cas9 and sgRNA constructs were delivered simultaneously into callus cells through bombardment with DNA-coated gold particles. Prior studies had demonstrated inefficient HDR-mediated gene replacement when *Agrobacterium* delivery of template DNA was utilized.[35,63] This may be attributable to the low quantity of donor DNA available for DSB repair through HDR when *Agrobacterium*-mediated transformation is used. Rice has also been used to demonstrate the feasibility of multiplex gene editing using multiple distinct guide RNAs that are capable of simultaneously forming ribonucleoprotein complexes with a common Cas9 protein within the same cell. The multiple sgRNAs can each be produced from a single polymerase III promoter (e.g., U6 or U3 promoter) or collectively be processed into multiple individual sgRNAs from a single transcript transcribed from a single U6 or U3 promoter.[64–66]

CRISPR/Cas9 has been shown to be efficient in generating knockout of both alleles of a single rice gene or multiple genes—or deletion of a gene cluster from a genomic region—even in T0 generation plants.[66–68] Such robust capacities have enabled scientists to assign gene function, dissect complex networks, and incorporate naturally occurring genetic variations into elite rice varieties in a timely, cost saving manner. For example, simple CRISPR gene knockout of *OsSWEET13*, a sucrose efflux transporter gene targeted by the TALE PthXa2 of *X. oryzae* to cause bacterial blight, resulted in plants that were completely resistant to the PthXo2-dependent Xoo strains—a resistance that is due to loss of function of the disease susceptibility gene (*OsSWEET13*) to Xoo pathogens.[69] CRISPR-modified rice lines have been used to reassess genes involved in rice yield,[70,71] to identify individual gene function,[72–75] to drive grain quality improvement,[76] and to accelerate hybrid rice breeding.[77,78]

5. USE OF CRISPR/CPF1 FOR GENE EDITING IN RICE

Despite the popularity of CRISPR/Cas9 as a tool for genome engineering and altered gene regulation, Cpf1 (CRISPR from *Prevotella* and *Francisella* 1—which belongs to the same class 2 CRISPR–Cas system as Cas9)[79] has emerged as an alternative, potentially promising, genome engineering tool.[80–82] Cpf1 is similar to Cas9 as defined by their RuvC nuclease domains, but it differs from Cas9 in several ways,[83] also as illustrated in Fig. 1C (compare model in B vs model in C). First, Cpf1 does not have the second HNH nuclease domain of Cas9 that is capable of cleaving the second DNA strand in a target. Instead, the capacity for DSB may be attributable to a putative novel nuclease cleavage domain.[84] Second, the CRISPR/Cpf1 system, unlike CRISPR/Cas9, lacks an obvious tracrRNA, and uses a single crRNA. Third and most importantly, CRISPR/Cpf1 uses a 5′-TTTN-3′ PAM (protospace adjacent motif) located at the 5′ end of target (or protospace) sequence and cleaves the target strand at 23 nt and the non-target strand at 18 nt downstream of the PAM sequence, resulting in a staggered cut with two 5-nt overhangs. In contrast, Cas9 uses a 5′-NGG-3′ PAM sequence at the 3′ end of target and causes a blunt-end cut 3-nt upstream of PAM[84–86] (Fig. 1C). Given the aforementioned characteristics of CRISPR/Cas9, CRISPR/Cpf1 provides specific advantages over Cas9 (in addition to being an alternate approach to Cas9, in general). For example, Cpf1-induced mutagenic insertions/deletions (indels) are relatively large (vs 1-bp indels in the majority of Cas9-induced mutations).[87] This attribute may be useful in functional analysis of regulatory DNA elements, noncoding (e.g., miRNA) genes, and is especially suitable for target sites in AT-rich regions. The staggered breakages by CRISPR/Cpf1 may also be more efficient for knockin of genetic elements in vivo.

Like Cas9 proteins, Cpf1 comes in different forms depending on the microbes from which they are derived, including FnCpf1 (from *Francisella novicida*), AsCpf1 (*Acidaminococcus* sp.), and LbCpf1 (*Lachnospiraceae bacterium*). Each are associated with their own unique cognate crRNAs—crRNAs that are similar but not identical in structure and ribonucleic acid sequence.[82] Initial success in the use of Cpf1 for targeted mutagenesis was demonstrated in human cultured cells,[82,88] mice,[89–91] plants,[92–96] and in cyanobacteria.[97] These examples represent just beginning of realizing the full potential of this promising system. Zetsche *et al.* were the first to use the programmable Cpf1 systems for genome editing and first demonstrated the feasibility of targeted mutagenesis with AsCpf1 and LbCpf1 in human cells at frequencies comparable to SpCas9.[82] By targeting genes in fertilized mouse eggs by microinjecting crRNAs and the corresponding mRNAs encoding AsCpf1 and LbCpf1, Kim *et al.* succeeded in obtaining mutant founder mice at high frequencies and demonstrated germline transmission of mutant alleles.[90]

Rice, again as an excellent model, has been used by multiple groups to demonstrate the feasibility of developing CRISPR/Cpf1 systems for genome editing in plants. Targeted mutagenesis using CRISPR/Cpf1 in rice was demonstrated among those research groups at varying efficiencies; the difference was probably due to different expression levels of crRNA and *Cpf1* genes and choice of the targets they used.[92,93,95,96]. For example, Xu *et al.* in the first publication describing use of the CRISPR/Cpf1 in plants demonstrated the feasibility of targeted mutagenesis with LbCpf1. Two endogenous rice genes (*OsPDS* and *OsBEL*) were targeted using only crRNAs containing the pre-crRNA architectures—not the mature 19-nt crRNA architecture. The mutation rates ranged from 13.6% to 41.2% with higher rates achieved using the full-length repeat/spacer/full-length repeat crRNA architecture (pre-crRNA type II) than with using the full-length repeat/spacer crRNA architecture (pre-crRNA type I). As expected, in both cases, the resulting indels were multinucleotide mutations. The authors also provided evidence for the heritability of the mutations they obtained.[96] In contrast, when the mature crRNA scaffold of CRISPR/FnCpf1 was used with tobacco and rice, mutations at even higher frequencies were detected at the target genomic loci.[92] For example, >60% rice callus lines were found to contain site-specific mutations; while the T0 transgenic plants bore mutations at even higher frequencies (up to 90%).

Most recently, Tang *et al.* adopted the use of a Pol II promoter[98] and a double-ribozyme system[99] to express and mature the crRNAs, respectively, to drive the CRISPR/Cpf1 system in rice.[95] In this crRNA expression cassette, the crRNA (the 19-nt scaffold and programmable 23-nt spacer) is sandwiched by the hammerhead ribozyme at the 5′ end and the hepatitis delta virus ribozyme at the 3′ end, and the chimeric ribozyme/crRNA sequences are preceded by the maize ubiquitin promoter. When rice codon-optimized Cpf1 (AsCpf1 or LbCpf1) was expressed using the same ubiquitin promoter and tested with its cognate crRNA in rice protoplasts and stable T0 transgenic plants, high-efficiency mutagenesis was achieved. The LbCpf1 was superior to AsCpf1 in percentage of mutations obtained as assessed using a protoplast system. In transgenic plants, a 100% mutation frequency has been achieved using four constructs targeting three rice genes; the majority of mutations were biallelic.[95] The higher efficiency of this CRISPR/Cpf1 system compared with that described in previous reports may be attributable to the double-ribozyme system, which produces mature crRNAs containing the exact sequences of native crRNAs at both ends.

The full exploitation of the CRISPR/Cpf1 system for precise genomic modification and gene regulation will require more research. The presumably higher efficiency for insertion of homologous DNA elements at the staggered cut caused by crRNA/Cpf1 (compared to that obtained with the blunt-end cuts generated by CRISPR/Cas9) needs to be documented. Likewise, multiplex gene targeting with CRISPR/Cpf1 achieved either through polycistronic crRNA or through multiple single crRNAs has yet to be demonstrated in plant systems. Importantly, delivery of crRNA/Cpf1 components as a ribonucleoprotein complex has been successfully demonstrated to support precise genome editing in plant protoplasts.[94] However, it must be pointed out that this DNA-free mechanism of gene editing that has important implications for regulatory purpose await testing for practical applications in agriculture.

6. USE OF BASE EDITOR IN RICE

Genuine nuclease-based gene sequence editing or gene replacement is achieved by leveraging the abilities of programmable nucleases to introduce site-specific DSBs and exploiting the host cellular DSB repair processes in presence of repair DNA template. However, the efficiency of this process is usually low and thus challenging for routine use. To overcome such constrains, the recently developed Base Editor (BE) technology enables precise DNA base edits without DSBs or exogenous DNA templates. It depends on the use of a standard sgRNA and a modified Cas9 fusion protein. The fusion protein is a "dead" Cas9 (dCas9) (or, in later versions, a Cas9 nickase) fused with a cytidine deaminase that has capacity to convert cytidine to uridine. Use of this modified Cas9 fusion protein/sgRNA combination leads to a single base pair substitutions in eukaryotic genomes.[100,101] Among several cytidine deaminases tested, the rat cytidine deaminase APOBEC1 was found to be the most efficient in converting cytidine to uridine.[100] APOBEC1 was fused to the N-terminus of the catalytically inactivated Cas9 (dCas9), and the fusion protein (BE1, the first-generation Base Editor) led to 0.8%–7.7% base editing at six sites targeted in the human genome. Fusion of the uracil glycosylase inhibitor (UGI) to the C-terminus of BE1, resulted in the second Base Editor BE2, that improved the base-editing efficiency and suppressed unintended deletions. UGI inhibited the uracil-DNA glycosylase function to prevent uracil base excision repair, a cellular process to correct the deaminated cytidine (or uridine) caused by the Cas9–cytidine deaminase. Furthermore, a third-generation Base Editor (BE3) made by replacement of dCas9 with a Cas9 (D10A) nickase, resulted in 37% C to T conversion within the cytidine deamination window, a stretch of DNA located between the position 13 and 17 upstream of PAM (Fig. 1D). Remarkably, only about 1.1% of indels on average at the six tested loci was detected in human cells.[100] The Cas9 nickase from *Staphylococcus aureus* (SaCas9, a Cas9 with a 5′-NNGRRT-3′ PAM and smaller than SpCas9) was superior to SpCas9 nickase in editing six endogenous loci in the human genome.[102] These findings validated the generality of using various Cas9 homologs for base editing of cytidine when fused with cytidine deaminase. Remarkably, when a mutant APOBEC1 was used, the deamination window is narrowed to three bases. When sea lamprey cytidine deaminase (so-called activation-induced cytidine deaminase) was used in the Komor architecture (deaminase-Cas9D10A-UGI) and tested in yeast, high efficiency of base editing was also observed.[101] Using the Komor system, other researchers have successfully achieved high efficiencies of base editing in mice,[103] rice, wheat, and maize.[104–106]

The successful demonstration of genome editing by using cytidine deaminase coupled with CRISPR/Cas9 points to the prospect of expanding the genome-editing potential with other CRISPR system (e.g., CRISPR/Cpf1). Narrowing the deamination of cytidine at target sites and increasing the possibility of precisely targeting a specific DNA base pair by using a suite of Base Editors will help to expand the value of genome editing in basic science and applied research.

7. CONCLUDING REMARKS

Crop improvements or breeding novel agronomic traits made through traditional biotechnology approaches have been rather slow and expensive to develop, and they have focused largely on traits that can generate a large market share, such as herbicide resistance and certain insect resistance traits. Modern, innovative genome-editing techniques can greatly accelerate research progress and improve the precision and scope of genetic improvements. Moreover, such techniques have the potential to decrease regulatory burdens, at least in the United States. The numerous benefits associated with genome-editing technology, without doubt, will change the face of basic plant research and crop breeding. Genome editing enables discovery of gene function through targeted mutagenesis, leading to trait discovery via the formation of new gene variants. It also simplifies and expedites the introduction of novel traits and maintenance of complex traits through gene stacking. For producers, this means more products of higher quality with fewer costs, less labor, as well as new markets and opportunities. For consumers, this new technology will mean the abundant availability of more nutritious food at lower cost.

Rice has been repeatedly used as an important platform for the testing and adoption of TALEN and CRISPR/Cas systems for genome editing—processes facilitated by efficient rice tissue culture and transformation technologies. However, rice genetic engineering by itself faces several constrains that limit application of genome-editing technologies. For example, most genome-editing studies were performed only with a limited number of transformable *japonica* rice varieties. It is still technically challenging to perform genetic engineering in the majority of rice cultivars, especially *indica* rice that remains recalcitrant to tissue culture and regeneration. Difficulty in rice protoplast culture and regeneration precludes the application of pre-assembled ribonucleoprotein of CRISPR/Cas or recombinant TALEN proteins for genome editing—processes that do not involve DNA in mutagenesis. Another issue is the uncertainty of regulatory rulings by governments in regard to genome edited rice. The stringency of such regulation will affect the application of genome-editing technologies in rice breeding and commercialization of TALEN- and CRISPR-modified rice. Overcoming these limitations will be a key to allowing full realization of the great potential of genome-editing technologies to meet the increasing challenge for achieving food security posed by the ever increasing world population.

REFERENCES

1. Bowers JE, Arias MA, Asher R, et al. Comparative physical mapping links conservation of microsynteny to chromosome structure and recombination in grasses. *Proc Natl Acad Sci USA*. 2005;102:13206–13211.
2. Yang Y, Li Y, Wu C. Genomic resources for functional analyses of the rice genome. *Curr Opin Plant Biol*. 2013;16:157–163.
3. Jiang Y, Cai Z, Xie W, Long T, Yu H, Zhang Q. Rice functional genomics research: progress and implications for crop genetic improvement. *Biotechnol Adv*. 2012;30:1059–1070.
4. Lo SF, Fan MJ, Hsing YI, et al. Genetic resources offer efficient tools for rice functional genomics research. *Plant Cell Environ*. 2016;39:998–1013.
5. Weeks D, Spalding M, Yang B. Use of designer nucleases for targeted gene and genome editing in plants. *Plant Biotechnol J*. 2016;14:483–495.
6. Puchta H. The repair of double-strand breaks in plants: mechanisms and consequences for genome evolution. *J Exp Bot*. 2005;56:1–14.
7. Lieber MR. The mechanism of double-strand DNA break repair by the non-homologous DNA end-joining pathway. *Annu Rev Biochem*. 2010;79:181–211.
8. Cottarel J, Frit P, Bombarde O, et al. A noncatalytic function of the ligation complex during nonhomologous end joining. *J Cell Biol*. 2013;200:173–186.
9. Gottlieb TM, Jackson SP. The DNA-dependent protein kinase: requirement for DNA ends and association with Ku antigen. *Cell*. 1993;72:131–142.
10. Mladenov E, Iliakis G. Induction and repair of DNA double strand breaks: the increasing spectrum of non-homologous end joining pathways. *Mutat Res*. 2011;711:61–72.
11. Knoll A, Fauser F, Puchta H. DNA recombination in somatic plant cells: mechanisms and evolutionary consequences. *Chromosome Res*. 2014;22:191–201.
12. Boch J, Bonas U, VanAlfen N, Bruening G, Leach J. Xanthomonas AvrBs3 family-type III effectors: discovery and function. *Annu Rev Phytopathol*. 2010;48(48):419–436.
13. White FF, Potnis N, Jones JB, Koebnik R. The type III effectors of Xanthomonas. *Mol Plant Pathol*. 2009;10:749–766.
14. Boch J, Scholze H, Schornack S, et al. Breaking the code of DNA binding specificity of TAL-type III effectors. *Science*. 2009;326:1509–1512.
15. Moscou MJ, Bogdanove AJ. A simple cipher governs DNA recognition by TAL effectors. *Science*. 2009;326:1501.
16. Deng D, Yan C, Pan X, et al. Structural basis for sequence-specific recognition of DNA by TAL effectors. *Science*. 2012;335:720–723.
17. Mak AN, Bradley P, Cernadas RA, Bogdanove AJ, Stoddard BL. The crystal structure of TAL effector PthXo1 bound to its DNA target. *Science*. 2012;335:716–719.

18. Li T, Huang S, Zhao X, et al. Modularly assembled designer TAL effector nucleases for targeted gene knockout and gene replacement in eukaryotes. *Nucleic Acids Res*. 2011;39:6315–6325.
19. Zhang F, Cong L, Lodato S, Kosuri S, Church GM, Arlotta P. Efficient construction of sequence-specific TAL effectors for modulating mammalian transcription. *Nat Biotechnol*. 2011;29:149–153.
20. Bogdanove AJ, Voytas DF. TAL effectors: customizable proteins for DNA targeting. *Science*. 2011;333:1843–1846.
21. Doyle E, Stoddard B, Voytas D, Bogdanove A. TAL effectors: highly adaptable phytobacterial virulence factors and readily engineered DNA-targeting proteins. *Trends Cell Biol*. 2013;23:390–398.
22. Wright DA, Li T, Yang B, Spalding MH. TALEN-mediated genome editing: prospects and perspectives. *Biochem J*. 2014;462:15–24.
23. Christian M, Cermak T, Doyle EL, et al. Targeting DNA double-strand breaks with TAL effector nucleases. *Genetics*. 2010;186:757–761.
24. Li T, Huang S, Jiang WZ, et al. TAL nucleases (TALNs): hybrid proteins composed of TAL effectors and FokI DNA-cleavage domain. *Nucleic Acids Res*. 2011;39:359–372.
25. Cermak T, Starker CG, Voytas DF. Efficient design and assembly of custom TALENs using the golden gate platform. *Methods Mol Biol*. 2015;1239:133–159.
26. Reyon D, Tsai SQ, Khayter C, Foden JA, Sander JD, Joung JK. FLASH assembly of TALENs for high-throughput genome editing. *Nat Biotechnol*. 2012;30:460–465.
27. Sakuma T, Ochiai H, Kaneko T, et al. Repeating pattern of non-RVD variations in DNA-binding modules enhances TALEN activity. *Sci Rep*. 2013;3:3379.
28. Schmid-Burgk JL, Schmidt T, Hornung V. Ligation-independent cloning (LIC) assembly of TALEN genes. *Methods Mol Biol*. 2015;1239:161–169.
29. Li T, Liu B, Spalding MH, Weeks DP, Yang B. High-efficiency TALEN-based gene editing produces disease-resistant rice. *Nat Biotechnol*. 2012;30:390–392.
30. Antony G, Zhou J, Huang S, et al. Rice xa13 recessive resistance to bacterial blight is defeated by induction of the disease susceptibility gene Os-11N3. *Plant Cell*. 2010;22:3864–3876.
31. Blanvillain-Baufumé S, Reschke M, Solé M, et al. Targeted promoter editing for rice resistance to Xanthomonas oryzae pv. Oryzae reveals differential activities for SWEET14-inducing TAL effectors. *Plant Biotechnol J*. 2017;15:306–317.
32. Ma L, Zhu F, Li Z, et al. TALEN-based mutagenesis of lipoxygenase LOX3 enhances the storage tolerance of rice (*Oryza sativa*) seeds. *PLoS One*. 2015;10, e0143877.
33. Shan Q, Zhang Y, Chen K, Zhang K, Gao C. Creation of fragrant rice by targeted knockout of the OsBADH2 gene using TALEN technology. *Plant Biotechnol J*. 2015;13:791–800.
34. Miki D, Zhu P, Zhang W, et al. Efficient generation of diRNAs requires components in the posttranscriptional gene silencing pathway. *Sci Rep*. 2017;7:301.
35. Li T, Liu B, Chen CY, Yang B. TALEN-mediated homologous recombination produces site-directed DNA base change and herbicide-resistant rice. *J Genet Genomics*. 2016;43:297–305.
36. Wiedenheft B, Sternberg SH, Doudna JA. RNA-guided genetic silencing systems in bacteria and archaea. *Nature*. 2012;482:331–338.
37. Ishino Y, Shinagawa H, Makino K, Amemura M, Nakata A. Nucleotide sequence of the iap gene, responsible for alkaline phosphatase isozyme conversion in *Escherichia coli*, and identification of the gene product. *J Bacteriol*. 1987;169:5429–5433.
38. Mojica FJ, Díez-Villaseñor C, Soria E, Juez G. Biological significance of a family of regularly spaced repeats in the genomes of archaea, bacteria and mitochondria. *Mol Microbiol*. 2000;36:244–246.

39. Mojica FJ, Ferrer C, Juez G, Rodríguez-Valera F. Long stretches of short tandem repeats are present in the largest replicons of the archaea Haloferax mediterranei and Haloferax volcanii and could be involved in replicon partitioning. *Mol Microbiol.* 1995;17:85–93.
40. Jansen R, Embden JD, Gaastra W, Schouls LM. Identification of genes that are associated with DNA repeats in prokaryotes. *Mol Microbiol.* 2002;43:1565–1575.
41. Makarova KS, Grishin NV, Shabalina SA, Wolf YI, Koonin EV. A putative RNA-interference-based immune system in prokaryotes: computational analysis of the predicted enzymatic machinery, functional analogies with eukaryotic RNAi, and hypothetical mechanisms of action. *Biol Direct.* 2006;1:7.
42. Barrangou R, Fremaux C, Deveau H, et al. CRISPR provides acquired resistance against viruses in prokaryotes. *Science.* 2007;315:1709–1712.
43. Garneau JE, Dupuis M, Villion M, et al. The CRISPR/Cas bacterial immune system cleaves bacteriophage and plasmid DNA. *Nature.* 2010;468:67–71.
44. Marraffini LA, Sontheimer EJ. CRISPR interference limits horizontal gene transfer in staphylococci by targeting DNA. *Science.* 2008;322:1843–1845.
45. Jinek M, Chylinski K, Fonfara I, Hauer M, Doudna JA, Charpentier E. A programmable dual-RNA-guided DNA endonuclease in adaptive bacterial immunity. *Science.* 2012;337:816–821.
46. Cong L, Ran F, Cox D, et al. Multiplex genome engineering using CRISPR/Cas systems. *Science.* 2013;339:819–823.
47. Mali P, Yang L, Esvelt KM, et al. RNA-guided human genome engineering via Cas9. *Science.* 2013;339:823–826.
48. Feng Z, Zhang B, Ding W, et al. Efficient genome editing in plants using a CRISPR/Cas system. *Cell Res.* 2013;23:1229–1232.
49. Jiang W, Zhou H, Bi H, Fromm M, Yang B, Weeks DP. Demonstration of CRISPR/Cas9/sgRNA-mediated targeted gene modification in Arabidopsis, tobacco, sorghum and rice. *Nucleic Acids Res.* 2013;41, e188.
50. Li JF, Norville JE, Aach J, et al. Multiplex and homologous recombination-mediated genome editing in Arabidopsis and Nicotiana benthamiana using guide RNA and Cas9. *Nat Biotechnol.* 2013;31:688–691.
51. Miao J, Guo D, Zhang J, et al. Targeted mutagenesis in rice using CRISPR-Cas system. *Cell Res.* 2013;23:1233–1236.
52. Nekrasov V, Staskawicz B, Weigel D, Jones JD, Kamoun S. Targeted mutagenesis in the model plant Nicotiana benthamiana using Cas9 RNA-guided endonuclease. *Nat Biotechnol.* 2013;31:691–693.
53. Shan Q, Wang Y, Li J, et al. Targeted genome modification of crop plants using a CRISPR-Cas system. *Nat Biotechnol.* 2013;31:686–688.
54. Upadhyay SK, Kumar J, Alok A, Tuli R. RNA-guided genome editing for target gene mutations in wheat. *G3 (Bethesda).* 2013;3:2233–2238.
55. Xie K, Yang Y. RNA-guided genome editing in plants using a CRISPR-Cas system. *Mol Plant.* 2013;6:1975–1983.
56. Mao Y, Zhang H, Xu N, Zhang B, Gou F, Zhu JK. Application of the CRISPR-Cas system for efficient genome engineering in plants. *Mol Plant.* 2013;6:2008–2011.
57. Li J, Meng X, Zong Y, et al. Gene replacements and insertions in rice by intron targeting using CRISPR-Cas9. *Nat Plants.* 2016;2:16139.
58. Sun Y, Zhang X, Wu C, et al. Engineering herbicide-resistant rice plants through CRISPR/Cas9-mediated homologous recombination of acetolactate synthase. *Mol Plant.* 2016;9:628–631.
59. Mikami M, Toki S, Endo M. Parameters affecting frequency of CRISPR/Cas9 mediated targeted mutagenesis in rice. *Plant Cell Rep.* 2015;34:1807–1815.

60. Hu X, Wang C, Fu Y, Liu Q, Jiao X, Wang K. Expanding the range of CRISPR/Cas9 genome editing in rice. *Mol Plant*. 2016;9:943–945.
61. Kaya H, Mikami M, Endo A, Endo M, Toki S. Highly specific targeted mutagenesis in plants using *Staphylococcus aureus* Cas9. *Sci Rep*. 2016;6:26871.
62. Mikami M, Toki S, Endo M. Precision targeted mutagenesis via Cas9 paired nickases in rice. *Plant Cell Physiol*. 2016;57:1058–1068.
63. Endo M, Mikami M, Toki S. Biallelic gene targeting in rice. *Plant Physiol*. 2016;170:667–677.
64. Ma X, Zhang Q, Zhu Q, et al. A robust CRISPR/Cas9 system for convenient, high-efficiency multiplex genome editing in monocot and dicot plants. *Mol Plant*. 2015;8:1274–1284.
65. Xie K, Minkenberg B, Yang Y. Boosting CRISPR/Cas9 multiplex editing capability with the endogenous tRNA-processing system. *Proc Natl Acad Sci USA*. 2015;112:3570–3575.
66. Zhou H, Liu B, Weeks DP, Spalding MH, Yang B. Large chromosomal deletions and heritable small genetic changes induced by CRISPR/Cas9 in rice. *Nucleic Acids Res*. 2014;42:10903–10914.
67. Minkenberg B, Xie K, Yang Y. Discovery of rice essential genes by characterizing a CRISPR-edited mutation of closely related rice MAP kinase genes. *Plant J*. 2017;89:636–648.
68. Zhang H, Zhang J, Wei P, et al. The CRISPR/Cas9 system produces specific and homozygous targeted gene editing in rice in one generation. *Plant Biotechnol J*. 2014;12:797–807.
69. Zhou J, Peng Z, Long J, et al. Gene targeting by the TAL effector PthXo2 reveals cryptic resistance gene for bacterial blight of rice. *Plant J*. 2015;82:632–643.
70. Li M, Li X, Zhou Z, et al. Reassessment of the four yield-related genes Gn1a, DEP1, GS3, and IPA1 in rice using a CRISPR/Cas9 system. *Front Plant Sci*. 2016;7:377.
71. Xu R, Yang Y, Qin R, et al. Rapid improvement of grain weight via highly efficient CRISPR/Cas9-mediated multiplex genome editing in rice. *J Genet Genomics*. 2016;43:529–532.
72. Lee SK, Eom JS, Hwang SK, et al. Plastidic phosphoglucomutase and ADP-glucose pyrophosphorylase mutants impair starch synthesis in rice pollen grains and cause male sterility. *J Exp Bot*. 2016;67:5557–5569.
73. Liu L, Zheng C, Kuang B, Wei L, Yan L, Wang T. Receptor-like kinase RUPO interacts with potassium transporters to regulate pollen tube growth and integrity in rice. *PLoS Genet*. 2016;12, e1006085.
74. Yamauchi T, Yoshioka M, Fukazawa A, et al. An NADPH oxidase RBOH functions in rice roots during lysigenous aerenchyma formation under oxygen-deficient conditions. *Plant Cell*. 2017;29:775–790.
75. Yuan J, Chen S, Jiao W, et al. Both maternally and paternally imprinted genes regulate seed development in rice. *New Phytol*. 2017; http://dx.doi.org/10.1111/nph.14510 [Epub ahead of print].
76. Sun Y, Jiao G, Liu Z, et al. Generation of high-amylose rice through CRISPR/Cas9-mediated targeted mutagenesis of starch branching enzymes. *Front Plant Sci*. 2017;8:298.
77. Li Q, Zhang D, Chen M, et al. Development of japonica photo-sensitive genic male sterile rice lines by editing carbon starved anther using CRISPR/Cas9. *J Genet Genomics*. 2016;43:415–419.
78. Zhou H, He M, Li J, et al. Development of commercial thermo-sensitive genic male sterile rice accelerates hybrid rice breeding using the CRISPR/Cas9-mediated TMS5 editing system. *Sci Rep*. 2016;6:37395.
79. Schunder E, Rydzewski K, Grunow R, Heuner K. First indication for a functional CRISPR/Cas system in Francisella tularensis. *Int J Med Microbiol*. 2013;303:51–60.

80. Carroll D. Genome engineering with targetable nucleases. *Annu Rev Biochem*. 2014;83:409–439.
81. Komor AC, Badran AH, Liu DR. CRISPR-based technologies for the manipulation of eukaryotic genomes. *Cell*. 2017;168:20–36.
82. Zetsche B, Gootenberg JS, Abudayyeh OO, et al. Cpf1 is a single RNA-guided endonuclease of a class 2 CRISPR-Cas system. *Cell*. 2015;163:759–771.
83. Shmakov S, Smargon A, Scott D, et al. Diversity and evolution of class 2 CRISPR-Cas systems. *Nat Rev Microbiol*. 2017;15:169–182.
84. Yamano T, Nishimasu H, Zetsche B, et al. Crystal structure of Cpf1 in complex with guide RNA and target DNA. *Cell*. 2016;165:949–962.
85. Dong D, Ren K, Qiu X, et al. The crystal structure of Cpf1 in complex with CRISPR RNA. *Nature*. 2016;532:522–526.
86. Nishimasu H, Ran FA, Hsu PD, et al. Crystal structure of Cas9 in complex with guide RNA and target DNA. *Cell*. 2014;156:935–949.
87. Kim HK, Song M, Lee J, et al. In vivo high-throughput profiling of CRISPR-Cpf1 activity. *Nat Methods*. 2017;14:153–159.
88. Kleinstiver BP, Tsai SQ, Prew MS, et al. Genome-wide specificities of CRISPR-Cas Cpf1 nucleases in human cells. *Nat Biotechnol*. 2016;34:869–874.
89. Hur JK, Kim K, Been KW, et al. Targeted mutagenesis in mice by electroporation of Cpf1 ribonucleoproteins. *Nat Biotechnol*. 2016;34:807–808.
90. Kim Y, Cheong SA, Lee JG, et al. Generation of knockout mice by Cpf1-mediated gene targeting. *Nat Biotechnol*. 2016;34:808–810.
91. Watkins-Chow DE, Varshney GK, Garrett LJ, et al. Highly efficient Cpf1-mediated gene targeting in mice following high concentration pronuclear injection. *G3 (Bethesda)*. 2017;7:719–722.
92. Endo A, Masafumi M, Kaya H, Toki S. Efficient targeted mutagenesis of rice and tobacco genomes using Cpf1 from Francisella novicida. *Sci Rep*. 2016;6:38169.
93. Hu X, Wang C, Liu Q, Fu Y, Wang K. Targeted mutagenesis in rice using CRISPR-Cpf1 system. *J Genet Genomics*. 2017;44:71–73.
94. Kim H, Kim ST, Ryu J, Kang BC, Kim JS, Kim SG. CRISPR/Cpf1-mediated DNA-free plant genome editing. *Nat Commun*. 2017;8:14406.
95. Tang X, Lowder LG, Zhang T, et al. A CRISPR-Cpf1 system for efficient genome editing and transcriptional repression in plants. *Nat Plants*. 2017;3:17018.
96. Xu R, Qin R, Li H, et al. Generation of targeted mutant rice using a CRISPR-Cpf1 system. *Plant Biotechnol J*. 2016. http://dx.doi.org/10.1111/pbi.12669.
97. Ungerer J, Pakrasi HB. Cpf1 is a versatile tool for CRISPR genome editing across diverse species of cyanobacteria. *Sci Rep*. 2016;6:39681.
98. Christensen AH, Quail PH. Ubiquitin promoter-based vectors for high-level expression of selectable and/or screenable marker genes in monocotyledonous plants. *Transgenic Res*. 1996;5:213–218.
99. Gao Y, Zhao Y. Self-processing of ribozyme-flanked RNAs into guide RNAs in vitro and in vivo for CRISPR-mediated genome editing. *J Integr Plant Biol*. 2014 56: 343–349.
100. Komor AC, Kim YB, Packer MS, Zuris JA, Liu DR. Programmable editing of a target base in genomic DNA without double-stranded DNA cleavage. *Nature*. 2016;533: 420–424.
101. Nishida K, Arazoe T, Yachie N, et al. Targeted nucleotide editing using hybrid prokaryotic and vertebrate adaptive immune systems. *Science*. 2016;353, aaf8729.
102. Kim YB, Komor AC, Levy JM, Packer MS, Zhao KT, Liu DR. Increasing the genome-targeting scope and precision of base editing with engineered Cas9-cytidine deaminase fusions. *Nat Biotechnol*. 2017;35:371–376. http://dx.doi.org/10.1038/nbt.3803.

103. Kim K, Ryu SM, Kim ST, et al. Highly efficient RNA-guided base editing in mouse embryos. *Nat Biotechnol.* 2017;35:435–437. http://dx.doi.org/10.1038/nbt.3816.
104. Li J, Sun Y, Du J, Zhao Y, Xia L. Generation of targeted point mutations in rice by a modified CRISPR/Cas9 system. *Mol Plant.* 2016;3:526–529.
105. Lu Y, Zhu JK. Precise editing of a target base in the rice genome using a modified CRISPR/Cas9 system. *Mol Plant.* 2016;10:523–525.
106. Zong Y, Wang Y, Li C, et al. Precise base editing in rice, wheat and maize with a Cas9-cytidine deaminase fusion. *Nat Biotechnol.* 2017;35:438–440. http://dx.doi.org/10.1038/nbt.3811.

CHAPTER SIX

Genome Editing to Improve Abiotic Stress Responses in Plants

Yuriko Osakabe[1], Keishi Osakabe
Faculty of Bioscience and Bioindustry, Tokushima University, Tokushima, Japan
[1]Corresponding author: e-mail address: osakabe.yuriko@tokushima-u.ac.jp

Contents

Abstract

Targeted modification of specific genes via genome editing is now used routinely to modify plant genomes. In developing new mutations in plant genomes using the widely used CRISPR/Cas9 system, it is important for further use in plant molecular studies and crop breeding that the mutations generated are heritable. To date, several improvements to increase efficiency and specificity have been developed to generate heritable mutations in various plant species. In this chapter, we focus on strategies to improve genome editing technology to increase heritability in plants, and summarize the process used to generate new mutant alleles of environmental stress response genes in plants. Such studies suggest further applications in molecular breeding to improve plant function using optimized plant CRISPR/Cas9 systems.

1. INTRODUCTION

Clustered regularly interspaced short palindromic repeats (CRISPR) and CRISPR-associated protein-9 nuclease (Cas9) have been utilized widely for genetic engineering in various organisms including plants.[1–7] In the CRISPR/Cas9 system, an RNA-guided nuclease, Cas9, from

Progress in Molecular Biology and Translational Science, Volume 149
ISSN 1877-1173
http://dx.doi.org/10.1016/bs.pmbts.2017.03.007

Streptococcus pyogenes, is engineered to target-specific DNA sequences using recognition via complementary base pairing to create a DNA double-stranded break (DSB). The protospacer adjacent motif (PAM) (5′-NGG-3′) is the recognition site of the target DNA sequence for the RNA-guided Cas9, forming a Cas9/gRNA/DNA complex. After induction of the DSB by gRNA-guided Cas9, DNA repair systems, comprising three major pathways [nonhomologous end joining (NHEJ), microhomology-mediated end joining (MMEJ), and homologous recombination (HR)], are activated at the DSB site. The NHEJ repair system creates insertion and/or deletion (indel) mutations, leading to frameshift-mediated gene disruption, whereas MMEJ is an error-prone end-joining using microhomologous sequences (5–25 bp) and results in deletion mutations. HR involves homologous DNA templates and generally results in precise repair, which can lead to gene replacement or insertion. The CRISPR/Cas9 system was developed in model plants such as *Arabidopsis*,[5,8-10] rice,[8-11] and tobacco,[5,6,10] but is now more widely utilized in other crop species, such as sorghum,[10] wheat,[12,13] maize,[14] soybean,[15,16] tomato,[17,18] and woody plants such as poplar[19,20] and apple.[21] To generate new mutant plant lines for molecular studies and crop breeding, an effective heritable mutation system has been developed. In addition, there has been much focus on the avoidance of DNA cleavage at nonspecific loci—so-called "off-target" mutations—in plant genome editing. Here, we summarize strategies to design and improve the CRISPR/Cas9 system with heritable high efficiency to generate new plant mutant lines with altered abiotic stress responses, and with low off-target effects. These studies will contribute to further applications in molecular breeding and biology.

2. OPTIMIZATION OF CRISPR/Cas9 SYSTEM TO GENERATE MUTANT PLANTS

2.1 Promoters for Cas9 Expression

The first reports of genome editing induced by CRISPR/Cas9 in plants appeared in 2013 and described transient DNA delivery systems for the introduction of CRISPR/Cas9 vectors in protoplasts of *Arabidopsis*,[5] rice,[11] and wheat,[11] or the Agroinfiltration method in *Nicotiana benthamiana* tobacco.[6] These studies targeted knockout of the phytone desaturase (*PDS*) gene—a key gene for carotenoid biosynthesis. The *PDS* gene is a good marker for plant genome editing in various plant species because an albino phenotype is generated when this gene is knocked out. The use of CRISPR/Cas9 to introduce heritable mutations in plants was first reported in *Arabidopsis*.[8,22] In these latter studies, *hCas9* gene (i.e., human codon optimized Cas9 gene) driven by the *CaMV35S* promoter and the gRNA expression cassette using the *AtU6-26* promoter were used to introduce mutations in *Arabidopsis*.[8] Heritable mutations, i.e., adhering to Mendel's law, were detected in the next generations. In addition, newly generated mutations were also detected, which were induced by CRISPR/Cas9 as the transgene transferred over generations. A study using the parsley ubiquitin promoter for driving *AtCas9* expression also showed efficient heritable mutations in *Arabidopsis*, with homozygotes obtained at high efficiency in the next generation.[23] These studies used constitutive promoters for *Cas9* expression, and enabled the effective introduction of mutations in *Arabidopsis*. In general, *Agrobacterium*-mediated in planta transformation has been used to introduce the CRISPR/Cas9 vector into *Arabidopsis*. Since the developing flower tissues include the gamete cells (pollen and ovule), and because embryos are used for transformation with *Agrobacterium* in this method, expression of *Cas9* must be sufficiently high in these tissues. Recently, tissue-specific promoters with high activity to express *Cas9* in the germline, and in embryo cells, have been used to generate heritable mutations in *Arabidopsis* with high efficiency (Table 1).[24–30] The *INCURVATA2* (*ICU2*) promoter (meristem-specific),[24] *EC1.2* promoter (embryo-specific),[25] *YAO* promoter (young tissue-specific),[26] *APETALA1* (*AP1*) promoter (flower-specific),[27] *SPOROCYTELESS* (*SPL*) promoter (germline-specific),[28] *Elongation Factor-1* (*EF1*) promoter (germline- and meristematic cell-specific),[29] and *RIBOSOMAL PROTEIN S5 A* (*RPS5A*) promoter (egg cell- and meristematic cell-specific)[30] have all now been used to induce efficient heritable mutation in *Arabidopsis* (Table 1 and Fig. 1). *Cas9* expression driven by the *AP1*, *SPL*, and *EF1* promoters induced biallelic and/or homo- and heteromutations in the T2 generation, although the mutation efficiency in the T1 generation was relatively low. We used the *EF1* promoter to produce the new allele for the abiotic stress-responsive gene, *OST2/AHA1*, in *Arabidopsis*. We will summarize this study in a later section of this review. In contrast to *Cas9* expression, the various promoters listed above have not been used to express gRNA. Instead, species-specific Pol III promoters have been used for gRNA expression; for example, the *AtU6–26* promoter has been used widely to induce mutation effectively in dicots.

Table 1 CRISPR/Cas9 for Heritable Genome Editing in *Arabidopsis*

Promoter	Expression	Cas9	gRNA Promoter	Target Gene	Mutation Frequency in T1	Mutation Frequency in T2	Refs.
INCURVATA2 (*ICU2*)	Meristem	hSpCas9	AtU6-26	*FT*	10%–85%	Homo-, hetero-, biallelic	24
EC1.2	Egg cell	Mayze zCas9	AtU6-26	*ETC2*, *TRY*, *CPC*	(multiple over 8.3% double and triple)	Homo-, Hetero-, biallelic	25
YAO	Cell proliferetaion stage	hSpCas9	AtU6-26	*BRI1*	90%	Homo-, hetero-, biallelic	26
APETALA1 (*AP1*)	Developing flower	hSpCas9	AtU6-26	*ABP1*	ND*	Homo-, Hetero-, biallelic	27
SPOROCYTELESS (*SPL*)	Germiline	hSPCas9	AtU6-26	*AP1*, *TT4*	10%–30%	Homo-, hetero-, biallelic	28
Elongation Factor-1 (*EF1*)	Meristematic and germline	coCas9	AtU6-1	*AtOST2*	Over 20%	Homo-, hetero-, biallelic	29
RPS5A	Constitutive (egg cells to meristematic)	hSpCas9	AtU6-26	*PDS3*, *AGAMOUS*, *DU01*	67%–85%	Homo-, hetero-, biallelic	30

*Not determined.

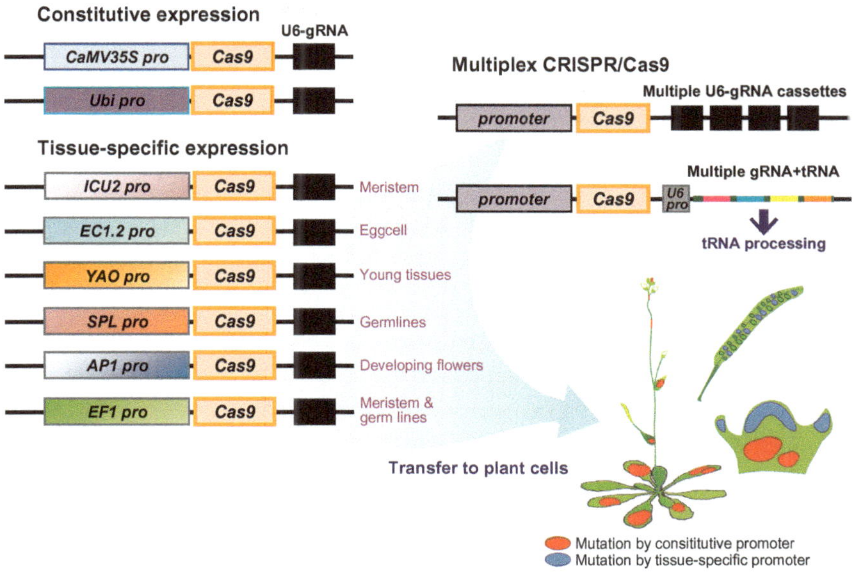

Fig. 1 Schematic illustrating CRISPR/Cas9 vectors using various types of promoter for Cas9 expression. Tissue-specific promoters induce *Arabidopsis* heritable mutations effectively. Multiplex CRISPR/Cas9 vectors are able to introduce and express multiple gRNAs.

2.2 gRNA Design

In the CRISPR/Cas9 system, one of the most important rate-limiting factors for inducing efficient mutation is selection of the 20 bp sequence and associated PAM sequence of the target gene for design and construction of the gRNA. Together with in silico tools for off-target searches, computational gRNA selection to predict the knockout efficiency of selected (on-target) gRNAs have been developed using large-scale guide RNA efficiency data sets from mammalian cells.[31–33] Similar screens to determine sgRNA on-target efficiency have also reported based on prediction of sgRNA stability, which depends on guanine enrichment and adenine depletion.[34] Recently, Doench *et al.* further refined the rules for sgRNA design using a pooled screen analysis of 9914 gRNAs covering all single mismatch and single RNA indels for 65 DNA targets.[35] Such on-target calculations now provide a bank of available in silico tools for gRNA design.

In parallel, to eliminate off-target effects and improve specificity, several other strategies have been studied, including: (1) the use of a mutated version of Cas9 (D10A, Cas9 nickase; Cas9n),[36] (2) careful design using computational gRNA selection, and (3) gRNAs truncated at the 5′ end (17–18 nucleotides)[37] (tru-gRNAs). Off-target efficiency in plant genome editing has been evaluated using Cas9n.[38,39] In this system, off-targets can be avoided by using two adjacent sgRNA target sequences to guide the Cas9n to generate a DSB at the target locus. tru-gRNAs have also been used to produce site-directed modifications in plants with no off-target effects.[29]

In many plants where cell proliferation and tissue culture are used to regenerate mature plants due to difficulty in crossbreeding, the generation of 100% mutant plant lines is important, not only to establish lines without mosaic mutation with wild type but also to purify homozygous and/or biallelic mutant lines for transformation with CRISPR/Cas9. Interestingly, recent studies have suggested that off-target mutations were detected less frequently than predicted in mature mutant plants. It has been reported that optimization of gRNA and Cas9 expression levels can increase both activity and specificity.[23,24,27] Thus, as described earlier, both optimization of gRNA and Cas9 expression levels, and ensuring tissue-specific Cas9 expression are necessary when developing in planta transformation systems to generate specific mutations effectively. In order to apply the CRISPR/Cas9 system to a variety of plant species, further studies will still be needed to optimize both an increase in efficient mutation rates and elimination of off-target effects.

3. DEVELOPING A NEW ALLELE OF A PLANT ABIOTIC STRESS-RESPONSIVE GENE

Adverse environmental factors, such as water deficiency, high salinity, heat, and nutrient stresses, are major factors limiting plant growth and crop productivity worldwilde,[40] and, therefore, abiotic stress-responsive factors are one of the most important targets for agricultural application of gene editing in plants. We focused on the gene encoding OPEN STOMATA 2 (OST2) (AHA1)—a plasma membrane H^+-ATPase in the stomata response in *Arabidopsis*[41]—and have successfully modified the *Arabidopsis* genome using CRISPR/Cas9 (Fig. 2).[29] Plasma membrane proton (H^+)-ATPases function to generate proton gradients in plant cells, and activate various transporters for the uptake of ions and metabolites.[40,41] The *Arabidopsis* genome harbors 11 members of the plasma membrane H^+-ATPase family, AHA1–AHA11,[42] which are composed of cytoplasmic N-terminal and C-terminal domains and a transmembrane domain including phosphorylation and nucleotide-binding sites.[43] The AHA C-terminal region is the major regulatory domain involved in inhibition of H^+-ATPase, and its

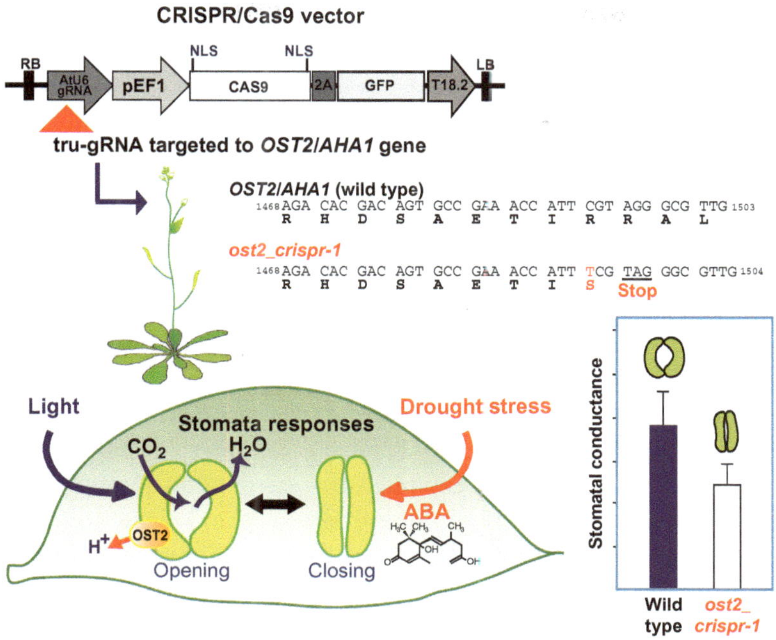

Fig. 2 Generation of a new allele of the *ost2* mutant using the tissue-specific *AtEF1* promoter for Cas9 expression. The newly generated mutant *Arabidopsis*, *ost2_crispr-1*, exhibited enhanced stomatal responses.

activation by phosphorylation and subsequent interaction with 14-3-3 proteins[44] also controls AHA function in plant cells.

We optimized the CRISPR/Cas9 system using coexpression of Cas9 and green fluorescent protein (GFP) via the 2A-coupled system (that allows expression of two coding regions from a gene containing a single promoter[45]) to monitor and select Cas9-expressing cells. We also used a tru-gRNA to eliminate off-target effects and to produce novel alleles of *OST2/AHA1* in *Arabidopsis*.[29] The 2A-coupled system to coexpress Cas9 and GFP in the same plant cell is very effective in selecting plants with high nuclease levels. Fine-tuning of adequate expression levels of both Cas9 and gRNAs together, which allows high mutation rates in plant cells, is needed to improve the CRISPR/Cas9 system for plant cells, thus monitoring of Cas9 expression with GFP is useful system in various plant species. In our study, we designed 17–18 nt (nucleotide) tru-gRNAs for stress-responsive genes, and the 18 nt tru-gRNA was found to induce mutation effectively in plant cells. The mutation rates varied with each of the tru-gRNAs and seemed to be affected strongly by the target sequence, as is also the case with

20 nt gRNAs.[29] Using the *CaMV35S* promoter, mutation rates in the *OST2/AHA1* locus in the T1 generation of transgenic *Arabidopsis* plants were up to 32.8% on average, with no off-target effects (evaluated using the amplicon sequence analyzed by next-generation sequencing). These results suggested that tru-gRNAs could be used effectively in the further application of plant genome editing, especially in important crop species.

To increase the levels of heritable mutation and homozygotes in the T2 generation in *Arabidopsis*, the *EF1* promoter—a strong tissue-specific promoter in meristematic tissues and the germline—was used to control *Cas9* expression in *Arabidopsis*. tru-gRNA Cas9 cassettes targeted for the *OST2/AHA1* gene and driven by the *EF1* promoter efficiently produced biallelic mutations in the T2 generation of *Arabidopsis*. Finally, a new mutant allele for *OST2/AHA1* (*ost2_crispr-1*) in *Arabidopsis* was obtained.[29] These *ost2_crispr-1 Arabidopsis* plants exhibited increased stomatal response, and higher sensitivity to the plant hormone abscisic acid and higher leaf temperature than the wild type, suggesting a rapid response of stomatal closing upon abiotic stress. The water-loss rates in leaves of the *ost2_crispr-1* were lower than those in the wild type due to the enhanced stomatal response.

In conclusion, we have developed an effective CRISPR/Cas9 system using tru-gRNAs and Cas9 driven by a tissue-specific promoter that can generate mutations effectively without off-target effects in plant genomes. The novel alleles for *OST2/AHA1* created by the optimized CRISPR/Cas9 exibited altered abiotic stress responses in *Arabidopsis*. Our results suggest further applications of CRISPR/Cas9-mediated genome editing to improve plant growth and stress responses. The strong advantages of this method likely can be applied to genome editing of many other crop species.

4. DEVELOPMENT OF NEW GENE DELIVERY SYSTEMS FOR CRISPR/Cas9

Genetic engineering is based on the introduction of transgenes and development of the resulting transgenic plants. Optimized and efficient gene delivery systems are thus required for plant genome editing. In addition, efficient regeneration systems and tissue culture to propagate transgenic plants also play a key role in these technologies. In plant genome editing, genetic transformation using an *Agrobacterium*-mediated system with plasmid DNA-harboring CRISPR/Cas9 has been developed, leading to high mutation efficiency. Recently, direct delivery methods with optimized PEG-mediated protoplast transfection of CRISPR/Cas9 ribonucleoproteins

(RNPs) or RNAs have been developed in several crops.[46–48] The methods reported by Woo *et al.* and Malnoy *et al.* utilize the direct delivery to protoplast cells of Cas9 protein and gRNA, with the mutated plant cells being regenerated to mature plants using tissue culture. These methods were limited to specific plants that utilize a regeneration system via tissue culture to obtain mature mutant plants; however, new direct delivery systems[46–48] where CRISPR/Cas9 without DNA being delivered are now opening up these techniques to allow next-generation plant molecular breeding. In addition to further studies on new technologies for plant genome editing giving highly efficient mutations and low off-target effects, research into the surrounding strategies such as tissue culture, micropropagation, and deliver systems is still needed to enhance the application of genome editing system to various species to support both plant genetic biology research and crop breeding.

ACKNOWLEDGMENTS

This work was supported by the Grant-in-Aid for Scientific Research C (26430190 to Y.O.) and Grant-in-Aid for Scientific Research on Innovative Areas from JSPS (15H01245 to Y.O.) from the Japan Society for the Promotion of Science (JSPS). This work was also supported by the Council for Science, Technology and Innovation (CSTI), Cross-ministerial Strategic Innovation Promotion Program (SIP), "Technologies for creating next-generation agriculture, forestry and fisheries" (funding agency: Bio-oriented Technology Research Advancement Institution, NARO), and NEDO.

REFERENCES

1. Cong L, Ran FA, Cox D, et al. Multiplex genome engineering using CRISPR/Cas systems. *Science*. 2013;339:819–823.
2. DiCarlo JE, Norville JE, Mali P, Rios X, Aach J, Church GM. Genome engineering in *Saccharomyces cerevisiae* using CRISPR-Cas systems. *Nucleic Acids Res*. 2013;41: 4336–4343.
3. Jinek M, Chylinski K, Fonfara I, Hauer M, Doudna JA, Charpentier E. RNA-programmed genome editing in human cells. *eLife*. 2013;2:e00471.
4. Mali P, Yang L, Esvelt KM, et al. RNA-guided human genome engineering via Cas9. *Science*. 2013;339:823–826.
5. Li JF, Norville JE, Aach J, et al. Multiplex and homologous recombination–mediated genome editing in *Arabidopsis* and *Nicotiana benthamiana* using guide RNA and Cas9. *Nat Biotechnol*. 2013;31:688–691.
6. Nekrasov V, Staskawicz B, Weigel D, Jones JD, Kamoun S. Targeted mutagenesis in the model plant Nicotiana benthamiana using Cas9 RNA-guided endonuclease. *Nat Biotechnol*. 2013;31:691–693.
7. Osakabe Y, Osakabe K. Genome editing with engineered nucleases in plants. *Plant Cell Physiol*. 2015;56:389–400.
8. Feng Z, Zhang B, Ding W, et al. Efficient genome editing in plants using a CRISPR/Cas9 system. *Cell Res*. 2013;23:1229–1232.

9. Mao Y, Zhang H, Xu N, Zhang B, Gou F, Zhu JK. Application of the CRISPR/Cas9 system for efficient genome engineering in plants. *Mol Plant.* 2013;6:2008–2011.
10. Jiang W, Zhou H, Bi H, Fromm M, Yang B, Weeks DP. Demonstration of CRISPR/Cas9/sgRNA-mediated targeted gene modification in Arabidopsis, tobacco, sorghum and rice. *Nucleic Acids Res.* 2013;41:e188.
11. Shan Q, Wang Y, Li J, et al. Targeted genome modification of crop plants using a CRISPR-Cas system. *Nat Biotechnol.* 2013;3(1):686–688.
12. Upadhyay SK, Kumar J, Alok A, Tuli R. RNA-guided genome editing for target gene mutations in wheat. *G3 Bethesda.* 2013;3:2233–2238.
13. Wang Y, Cheng X, Shan Q, et al. Simultaneous editing of three homoeoalleles in hexaploid bread wheat confers heritable resistance to powdery mildew. *Nat Biotechnol.* 2014;32:947–951.
14. Liang Z, Zhang K, Chen K, Gao C. Targeted mutagenesis in *Zea mays* using TALENs and the CRISPR/Cas9 system. *J Genet Genomics.* 2014;41:63–68.
15. Cai Y, Chen L, Liu X, et al. CRISPR/Cas9-mediated genome editing in soybean hairy roots. *PLoS One.* 2015;10:e0136064.
16. Jacobs TB, LaFayette PR, Schmitz RJ, Parrott WA. Targeted genome modifications in soybean with CRISPR/Cas9. *BMC Biotechnol.* 2015;15:16.
17. Ueta R, Abe C, Watanabe T, et al. Rapid breeding of parthenocarpic tomato plants using CRISPR/Cas9. *Sci Rep.* 2017;7:507.
18. Soyk S, Müller NA, Park SJ, et al. Variation in the flowering gene SELF PRUNING 5G promotes day-neutrality and early yield in tomato. *Nat Genet.* 2017;49:162.
19. Fan D, Liu T, Li C, et al. Efficient CRISPR/Cas9-mediated targeted mutagenesis in Populus in the first generation. *Sci Rep.* 2015;5:12217.
20. Zhou X, Jacobs TB, Xue LJ, Harding SA, Tsai CJ. Exploiting SNPs for biallelic CRISPR mutations in the outcrossing woody perennial Populus reveals 4-coumarate: CoA ligase specificity and redundancy. *New Phytol.* 2015;208:298–301.
21. Nishitani C, Hirai N, Komori S, et al. Efficient genome editing in apple using a CRISPR/Cas9 system. *Sci Rep.* 2016;6:31481.
22. Feng Z, Mao Y, Xu N, et al. Multigeneration analysis reveals the inheritance, specificity, and patterns of CRISPR/Cas-induced gene modifications in Arabidopsis. *Proc Natl Acad Sci USA.* 2014;111:4632–4637.
23. Fauser F, Schiml S, Puchta H. Both CRISPR/Cas-based nucleases and nickases can be used efficiently for genome engineering in *Arabidopsis thaliana. Plant J.* 2014;79:348–359.
24. Hyun Y, Kim J, Cho SW, Choi Y, Kim JS, Coupland G. Site-directed mutagenesis in *Arabidopsis thaliana* using dividing tissue-targeted RGEN of the CRISPR/Cas system to generate heritable null alleles. *Planta.* 2015;241:271–284.
25. Wang ZP, Xing HL, Dong L, et al. Egg cell-specific promoter-controlled CRISPR/Cas9 efficiently generates homozygous mutants for multiple target genes in Arabidopsis in a single generation. *Genome Biol.* 2015;16:144.
26. Yan L, Wei S, Wu Y, et al. High-efficiency genome editing in *Arabidopsis* using *YAO* promoter-driven CRISPR/Cas9 system. *Mol Plant.* 2015;8:1820–1823.
27. Gao Y, Zhang Y, Zhang D, Dai X, Estelle M, Zhao Y. Auxin binding protein 1 (ABP1) is not required for either auxin signaling or Arabidopsis development. *Proc Natl Acad Sci USA.* 2015;112:2275–2280.
28. Mao Y, Zhang Z, Feng Z, et al. Development of germ-line-specific CRISPR-Cas9 systems to improve the production of heritable gene modifications in *Arabidopsis. Plant Biotechnol J.* 2016;14:519–532.
29. Osakabe Y, Watanabe T, Sugano SS, et al. Optimization of CRISPR/Cas9 genome editing to modify abiotic stress responses in plants. *Sci Rep.* 2016;6:26685.

30. Tsutsui H, Higashiyama T. pKAMA-ITACHI vectors for highly efficient CRISPR/Cas9-mediated gene knockout in *Arabidopsis thaliana*. *Plant Cell Physiol*. 2017. http://dx.doi.org/10.1093/pcp/pcw191.
31. Doench JG, Hartenian E, Graham DB, et al. Rational design of highly active sgRNAs for CRISPR-Cas9-mediated gene inactivation. *Nat Biotechnol*. 2014;32:1262–1267.
32. Chari R, Mali P, Moosburner M, Church GM. Unraveling CRISPR-Cas9 genome engineering parameters via a library-on-library approach. *Nat Methods*. 2015;12:823–826.
33. Xu H, Xiao T, Chen CH, et al. Sequence determinants of improved CRISPR sgRNA design. *Genome Res*. 2015;25:1147–1157.
34. Moreno-Mateos MA, Vejnar CE, Beaudoin J-D, et al. CRISPRscan: designing highly efficient sgRNAs for CRISPR-Cas9 targeting in vivo. *Nat Methods*. 2015;12:982–988.
35. Doench JG, Fusi N, Sullender M, et al. Optimized sgRNA design to maximize activity and minimize off-target effects of CRISPR-Cas9. *Nat Biotechnol*. 2016;34:184–191.
36. Ran FA, Hsu PD, Lin CY, et al. Double nicking by RNA-guided CRISPR/Cas9 for enhanced genome editing specificity. *Cell*. 2013;154:1380–1389.
37. Fu Y, Sander JD, Reyon D, Cascio VM, Joung JK. Improving CRISPR-Cas nuclease specificity using truncated guide RNAs. *Nat Biotechnol*. 2014;32:279–284.
38. Fauser F, Schiml S, Puchta H. Both CRISPR/Cas-based nucleases and nickases can be used efficiently for genome engineering in *Arabidopsis thaliana*. *Plant J*. 2014;79:348–359.
39. Schiml S, Fauser F, Puchta H. The CRISPR/Cas system can be used as nuclease for in planta gene targeting and as paired nickases for directed mutagenesis in Arabidopsis resulting in heritable progeny. *Plant J*. 2014;80:1139–1150.
40. Osakabe Y, Osakabe K, Shinozaki K, Tran LS. Response of plants to water stress. *Front Plant Sci*. 2014;5:86.
41. Palmgren MG. Plant plasma membrane H+-ATPases: powerhouses for nutrient uptake. *Annu Rev Plant Physiol Mol Biol*. 2001;52:817–845.
42. Baxter I, Tchieu J, Sussman MR, et al. Genomic comparison of P-type ATPase ion pumps in Arabidopsis and rice. *Plant Physiol*. 2003;132:618–628.
43. Pedersen BP, Buch-Pedersen MJ, Morth JP, Palmgren MG, Nissen P. Crystal structure of the plasma membrane proton pump. *Nature*. 2007;450:1111–1114.
44. Svennelid F, Olsson A, Piotrowski M, et al. Phosphorylation of Thr-948 at the C terminus of the plasma membrane H+-ATPase creates a binding site for the regulatory 14-3-3 protein. *Plant Cell*. 1999;11:2379–2391.
45. Duda K, Lonowski LA, Kofoed-Nielsen M, et al. High-efficiency genome editing via 2A-coupled co-expression of fluorescent proteins and zinc finger nucleases or CRISPR/Cas9 nickase pairs. *Nucleic Acids Res*. 2014;42:e84
46. Woo JW, Kim J, Kwon SI, et al. DNA-free genome editing in plants with preassembled CRISPR-Cas9 ribonucleoproteins. *Nat Biotechnol*. 2015;33:1162.
47. Malnoy M, Viola R, Jung MH, et al. DNA-free genetically edited grapevine and apple protoplast using CRISPR/Cas9 ribonucleoproteins. *Front Plant Sci*. 2016;7:1904.
48. Zhang Y, Liang Z, Zong Y, et al. Efficient and transgene-free genome editing in wheat through transient expression of CRISPR/Cas9 DNA or RNA. *Nat Commun*. 2016;7:12617.

CHAPTER SEVEN

CRISPR/Cas9-Enabled Multiplex Genome Editing and Its Application

Bastian Minkenberg, Matthew Wheatley, Yinong Yang[1]
The Pennsylvania State University, University Park, PA, United States
[1]Corresponding author: e-mail address: yuy3@psu.edu

Contents

Abstract

The CRISPR/Cas9 system is a prevalent and versatile genome-editing tool of choice for basic and applied biological research. An exchange of a 20-bp spacer sequence in the gRNA can easily reprogram Cas9 to target a different DNA site. By expressing or providing multiple gRNAs, the system also enables multiplex genome editing at high efficiencies. Current approaches for providing multiple gRNAs in vivo include the use of multigene cassettes to express several gRNAs, Csy4-based excision, arrays of crRNAs, ribozyme-flanked gRNAs, tRNA-dependent cleavage of gRNAs, and direct introduction

Progress in Molecular Biology and Translational Science, Volume 149
ISSN 1877-1173
http://dx.doi.org/10.1016/bs.pmbts.2017.05.003

of Cas9 proteins preloaded with different gRNAs. By simultaneously targeting multiple DNA sequences, multiplex genome editing can be used to knockout multiple genes or delete chromosomal fragments. Off-target risk can also be reduced by Cas9-dimers that require the simultaneous expression of two gRNAs. With multiple gRNAs, specific gene expression or methylation status can be efficiently controlled by dCas9 fused to activators, repressors, methyltransferase, demethylase, or other functional domains. As a result, multiplex genome editing is expected to accelerate functional discovery of plant genes as well as genetic improvement of agricultural crops.

1. INTRODUCTION

Clustered regularly interspaced short palindromic repeats (CRISPR)/CRISPR-associated protein 9 (Cas9) became the most popular genome-editing tool during the last 5 years due to its simplicity and versatility compared to the previous genome-editing systems. The two most popular genome-editing systems before CRISPR/Cas9 was zinc-finger nucleases (ZFNs) and transcription activator-like effector nucleases (TALENs). ZFNs and TALENs rely on protein–DNA interactions to provide sequence specificity to their targets.[1] ZFNs recognize DNA with a series of zinc-finger DNA-binding domains that are fused to *Fok*I endonuclease domains.[2] The *Fok*I domains are dimerization-dependent and the two ZFN-targeting sites need to be close enough to allow *Fok*I dimerization. This in turn leads to the generation of a double-strand break through *Fok*I nuclease activity. A disadvantage of using ZFNs is that different zinc-finger domains show interdependence that makes it necessary to screen for combinations of domains that can reliably and effectively target a specific sequence.[3] Like ZFNs, TALENs also consist of DNA-binding domains coupled to dimerization-dependent *Fok*I domains. However, the DNA-binding domains follow a predictable code that allows for a more rational design of this site-specific nuclease.[4,5] Each domain consists of 34 amino acids and the amino acids in the 12th and 13th position determine domain-nucleotide specificity.[1] In contrast to ZFNs and TALENs which rely on protein–DNA interactions, CRISPR/Cas9 uses RNA–DNA base pairing. This specificity is provided by the single guide RNA (gRNA) to direct the Cas9 endonuclease to its complementary DNA target site. A simple exchange of a 20-bp spacer sequence in the gRNA can reprogram Cas9 to target a different protospacer sequence.[6,7]

One of the CRISPR/Cas9's biggest advantages is its ability for multiplex genome editing. It is only possible to use TALENs and ZFNs for multiplex genome editing when targeting a conserved sequence that is shared between target sites or via the introduction of multiple distinct paired nucleases. CRISPR/Cas9, on the other hand, can efficiently target multiple sites by simply providing several distinct gRNAs along with Cas9 protein.[8,9] The fact that an exchange of a 20-bp sequence within the gRNA is enough to change Cas9's target specificity makes CRISPR/Cas9 the optimal multiplexing tool compared to other available site-specific nucleases. In plants, efficient and simultaneous editing has been achieved at up to eight targeting sites.[9] CRISPR/Cas9-enabled multiplex genome editing is an exciting development that promises to accelerate basic research for plant biology as well as precision breeding for agricultural crops.

2. STRATEGIES TO ACHIEVE CRISPR/Cas9-ENABLED MULTIPLEX GENOME EDITING

Multiplex editing enables sophisticated applications for genome engineering, including multigene knockouts, chromosomal deletion and translocation, gene or promoter knock-in, gene activation and repression, or epigenome modifications, etc. Several research groups have developed different ways to efficiently express multiple gRNAs to take advantage of these enhanced editing techniques. In general, multiplex editing toolboxes are divided into systems that express multiple gRNAs with their own promoter and terminator, or systems that express multiple gRNAs via a single promoter and terminator. Each of these strategies comes with its own advantages and disadvantages, especially in regards to the ease of cloning, the number of gRNAs that can be expressed, and the efficiency of editing. One advantage of the methods that process a primary transcript to produce mature gRNAs is that they could be driven by a Pol II instead of a Pol III promoter to lift sequence restrictions or to allow for tissue-specific expression. Below is an overview of various methods to express multiple gRNAs for multiplex genome editing (also see Fig. 1).

2.1 Individual Expression Cassettes for Each gRNA

The most common approach for the production of multiple gRNAs is to express each individual gRNA by its own promoter and terminator as shown in Fig. 1A. Several laboratories have used this approach and generated

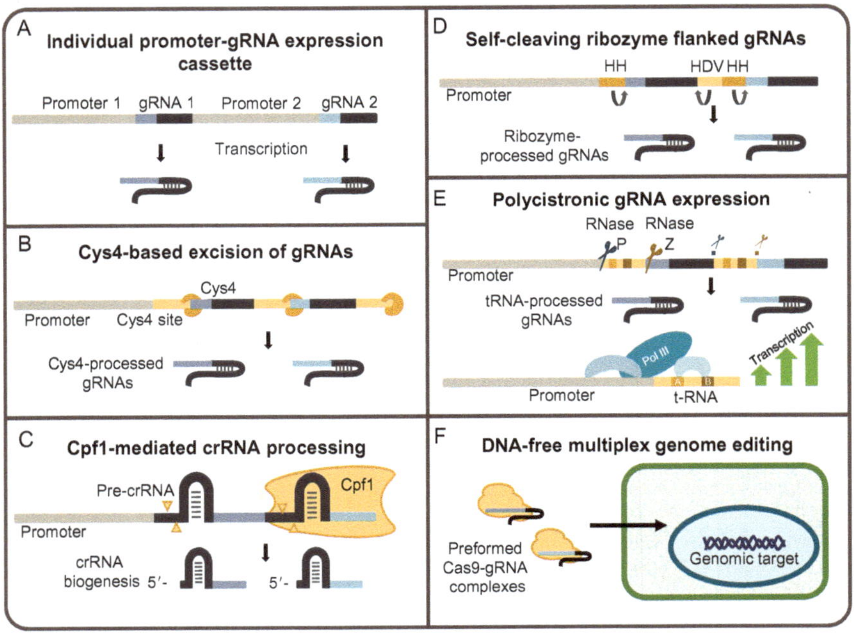

Fig. 1 Diverse strategies for multiplex gRNA expression.

toolboxes for expression of multiple gRNAs with these individual cassettes in plants.[10–21] The size of an expression cassette can vary between 500 and 800 bp depending on the promoter sequence used.[12,13,21,22] Since most plant transformation technologies preferably use only one vector, size limitations can be a concern when expressing an increased number of gRNAs. An additional challenge is balancing the trade-off between the assembled insert size and cloning efficiency.

A recent study using individual cassettes compared efficiencies of cloning an increasing number of cassette arrays based on a Golden Gate assembly method.[21] The assembly used a two-step method in which annealed oligonucleotides are inserted into a plasmid with a full expression cassette first. The oligonucleotides encode for distinct gRNA spacer sequences and carry an adapter at the ends to facilitate cloning into the vector by type-II restriction enzyme site digestions. Afterward, the Golden Gate reaction joins the individual cassettes to an array of multiple cassettes that can be transferred to an expression vector.[21] However, a satisfying cloning efficiency could be only achieved for up to five individual expression cassettes and the authors experienced great difficulties when they tried to join six to eight cassettes in one reaction. The limitation of cloning efficiency might be caused by the size of each individual cassette. Similarly, other publications only tested or reported assemblies of two to eight expression cassettes, indicating that this method quickly reaches limitations and is not suitable for expression of an increasing number of gRNAs.[10–21]

Interestingly, one study found a way to circumvent the previous problem of cloning efficiency by a method they called "stackable arrays."[23] The authors first created an array of up to four gRNA expression cassettes that were synthesized by a company. Multiple arrays were then cloned together to form a stacked array of gRNA expression cassettes. Using stackable arrays, up to 14 gRNA expression units could be cloned into one vector. This method enables cloning of a high number of cassettes but it requires a synthesis of parts by a company and may not be feasible for some researchers.

The development of multiplex expression cassettes has directly affected the amount of data generated regarding editing efficacy in different species. As not all studies provided data for multigene mutations or deletions, we compare here only the insertion and deletion frequencies for individual gRNAs. For *Arabidopsis*, reported editing efficiencies varied greatly with 10%–93% depending on the promoter used and the gRNA spacer sequence.[10,12,13,15,23] Efficiency reached 33.3%–90% in rice plants,[13,14,21] up to 60% in maize,[10] only 9.9%–12.1% in *Nicotiana benthamiana*,[11] 40.7%–52% in watermelon,[18] 60%–83% in potato,[24] 59.5%–68.7% in the moss *Scopelophila cataractae*, and 45%–67.8% in the model moss *Physcomitrella patens*.[25]

Considering the simplicity of the approach, expression of individual gRNA cassettes provides efficient editing. The low efficacy that was reported by some studies could be explained by different promoter or spacer sequences. Screening for better promoters and the choice of efficient spacer sequences could further improve the efficacy in such cases. However, the *Agrobacterium*-mediated transformation is the most popular transformation method for plants. Each *Agrobacterium* strain carries only one type of T-DNA plasmid that is transferred into the plant genome and therefore a more compact expression system for gRNAs would provide an advantage for plant transformation. Recently, researchers have developed several systems that can express multiple gRNAs or CRISPR locus RNAs (crRNAs) from a single expression cassette.

2.2 Csy4-Based Excision of gRNAs

One method to express multiple gRNAs from a single transcript is, like CRISPR/Cas9 itself, based on bacterial CRISPR systems. Csy4 is a *Pseudomonas aeruginosa* endoribonuclease needed for crRNA biogenesis of the

CRISPR subtype I—F.[26,27] Tsai *et al.* developed a system to use Csy4 for the excision of multiple gRNAs from an artificial locus.[28] Multiple gRNAs are fused together and divided by a Csy4 excision site at their 3′ end (Fig. 1B). When coexpressing *Csy4* with an artificial gRNA array, mature gRNAs are produced that can target multiple sites.[28] They used this technique to express multiple gRNAs for dimeric CRISPR RNA-guided *Fok*I nucleases. This system was developed for mammalian systems and reported to yield efficiencies of up to 90%.[28] The same expression system could also be used for other multiplex editing tasks. This method could easily be adapted for expression in plants by providing the right Pol III promoter. However, a major disadvantage is the requirement of expressing the foreign Csy4 endoribonuclease in addition to gRNA and Cas9. A system that does not need to express an additional nuclease might be more desirable.

2.3 Arrays to Express Multiple crRNAs for Cpf1

Another system for multiplex editing that was first developed in mammalian cells makes use of the CRISPR/Cpf1 system.[29] Like Cas9, Cpf1 is a member of the class II CRISPR system that only needs one enzyme for activity. One distinction is however that the CRISPR/Cpf1 system does not need trans-acting crRNAs (tracrRNAs).[30] Cutting specificity of this system is determined by crRNA arrays that are processed by Cpf1 into mature single crRNAs.[29,30] Therefore, Cpf1 itself can be used for crRNA processing and gene editing (Fig. 1C), which has the advantage that no additional nuclease needs to be expressed when compared to Csy4-dependent processing. This makes CRISPR/Cpf1 a facile multiplexing tool. Recent studies adapted the CRISPR/Cpf1 system successfully in plants but they did not test its multiplexing ability.[31,32] Summerized below are the efficiencies from these studies that may use different variants of Cpf1.

One study used a rice-codon optimized version of the *Lachnospiraceae bacterium* ND2006 Cpf1 (*Lb*Cpf1). They only achieved editing if a pre-crRNA was transcribed instead of mature crRNA.[31] The mutation frequency in T_0 rice plantlets was 13.6%–41.2%. Another study tested rice-codon and an *Arabidopsis*-codon optimized *Francisella novicida* Cpf1 (*Fn*Cpf1) in rice and tobacco, respectively. In contrast to the previously described study, the authors achieved editing by directly transcribing the crRNA from a U6 Pol III promoter.[32] The efficiency in tobacco and rice reached an average of 28.2% and 47.2%, respectively.[32] A third recent study tested the plant-codon optimized versions of *Acidaminococcus* sp. BV3L6

Cpf1 (*As*Cpf1) and *L. bacterium* ND2006 (*Lb*Cpf1).[33] Similar to the first study, they were not able to detect cleavage activity by directly expressing the mature crRNA from a U6 Pol III promoter. They overcame this obstacle by exact processing of the crRNA from the primary transcript by a method explained in the subsequent section. In short, they flanked the crRNA with two self-cleaving ribozymes and expressed the cassette with a maize ubiquitin Pol II promoter. They first tested both Cpf1 variants in rice protoplasts and found 0.6%–10% mutation frequency for *As*Cpf1 and 15%–25% mutation frequency for *Lb*Cpf1, concluding that mutation with *Lb*Cpf1 is more efficient in rice.[33] Testing of *Lb*Cpf1 in stable transgenic rice lines revealed an even higher mutation frequency of up to 100%. Even though none of the studies in plants so far tested the multiplexing ability of Cpf1, it seems that *Lb*Cpf1 and *Fn*Cpf1 are noteworthy alternatives to Cas9 in regard to multiplex genome engineering.

CRISPR/Cpf1 has two additional advantages because of the way the nuclease recognizes the cutting sites and the domains cut the DNA substrate. First, *Fn*Cpf1 recognize a 5′-TTN-3′ and *Lb*Cpf1 a 5′-TTTN-3′ protospacer adjacent motif (PAM) in contrast to the 5′-NGG-3′ PAM of the widely used *Streptococcus pyogenes* Cas9 (*Sp*Cas9).[32] This feature potentially increases the number of editable genome sites because this allows to target AT rich sites. Additionally, the total number of Cas9 NGG sites in *Arabidopsis* is 8,045,920 while the total number of TTN sites in the same genome is 3.5 times higher with 27,518,452 (calculated via EMBOSS fuzznuc pattern search and TAIR10 genome). *Fn*Cpf1 promises to increase the total number of sites that could be targeted. However, only a thorough bioinformatic evaluation of target specificity will clarify if the increased number of PAMs also translates into an increased number of specific targets. Second, Cpf1 produces a staggered cut with a 4–5 nt overhang.[30] This feature might be especially valuable for sophisticated genome-editing applications like gene stacking or tagging because the overhang enables the directional insertion of a gene or gene part.

2.4 Self-Cleaving Ribozyme Flanked gRNAs

Maturation of gRNAs by self-cleaving ribozymes is an approach that was first developed in yeast and could be exploited for multiplex genome editing.[34] The system takes advantage of the nuclease activity of modified ribozymes,[34] which are RNA molecules that catalyze reactions similar to enzymes. Gao and Zhao (2014) designed an artificial gene called a

Ribozyme-gRNA-Ribozyme gene (RGR). In an RGR, the gRNA is flanked by a Hammerhead-type (HH) ribozyme on its 5′-end and hepatitis delta virus (HDV) ribozyme on its 3′-end (Fig. 1D). In the initial study, this system was only shown to provide expression of one gRNA that gets matured by self-cleavage of the ribozymes. The authors suggested that this system could be easily modified into a multiplexed system by tandemly joining several RGR cassettes that could be expressed by only one promoter and terminator at the beginning and end, respectively.

Recent studies have used RGRs for multiplex genome editing in non-plant organisms. Two studies, one in zebrafish and one in *Leishmania donovani*, reported successful processing of two gRNAs using the RGR system.[35,36] Another study in human cell lines used RGRs to process up to four gRNAs to target two genes; however, they experienced a reduction in gRNA processing efficiency compared to only two gRNAs, indicating limitations of the system if higher numbers of gRNAs are expressed.[37] It is expected that the RGR system can be used for efficient expression of multiple gRNAs in plants.

2.5 tRNA-Dependent Expression and Processing of gRNAs by Endogenous RNases

Similar to the earlier method, a polycistron for multiple gRNA expression can be constructed out of other RNAs that are processed by cell's endogenous RNases. The polycistronic tRNA-gRNA strategy is a thoroughly tested approach for expression of multiple gRNAs from a single transcript.[9] Xie *et al.* developed an artificial gene construct that expresses multiple gRNAs by fusing a tRNA to a gRNA and tandemly arraying these in a polycistronic expression cassette (Fig. 1E). The tandem arrays are expressed by a single Pol III promoter and its terminator, and the endogenous RNaseZ and RNaseP recognize and cut out the tRNAs, thereby releasing mature gRNAs with different spacer sequences.[9] Every organism possesses RNases that process pre-tRNAs into mature tRNAs. Their endogenous RNases also recognize the cloverleaf structure of the tRNAs in the PTGs, excise them, and release multiple individual gRNAs.[9] Since the tRNA processing machinery is well conserved, other researcher have reported the effectiveness of PTGs in various plants such as wheat and maize,[38,39] as well as non-plant organisms like *Drosophila*, yeast, and human cells.[40–42]

In addition to successful processing of the primary transcript, a gRNA driven by Pol III promoter together with a tRNA was shown to have an increase of expression by 3–30-fold.[9] This boost in expression is probably caused by the A- and B-boxes in the tRNA sequences, which act as promoter/enhancer to recruit transcription factors (Fig. 1E).[43] The high expression of gRNAs also translates into higher editing efficiencies. While stable transformation of rice with the typical expression system achieved only up to 60% editing efficiency, rice transformed with the PTG system achieved up to 100% edited sites including 100% biallelic mutants in one case.[9,44] The original study in rice showed successful and simultaneous processing of up to eight mature gRNAs, while the study in wheat tested up to three gRNAs[38] and the study in maize up to four gRNAs in one array.[39] While the study in wheat did not test the construct in stable transformants,[38] the maize study showed a high editing efficiency of 85.7%–100% for stable maize mutants,[39] indicating that PTGs provide a very efficient way to coexpress several gRNAs from a single transcript and boost editing efficiencies. Port *et al.* showed in *Drosophila* that in addition to its application in CRISPR/Cas9, PTGs is also effective for expressing functional crRNAs for CRISPR/Cpf1-mediated genome editing.[40]

Aside from polycistrons being comprised of tRNA flanked gRNAs, another group has validated the functional multiplexing of tandem repeats of gRNAs-shRNA transcripts. The endogenous ribonuclease DROSHA will excise shRNAs in polycistronic transcripts releasing the gRNAs.[45] By utilizing this methodology genome editing and RNAi can simultaneously be achieved for advanced multiplex genome editing applications.[45] However, a combination of the tRNA and shRNA processing methods has yet to be tested, but holds promise due to the boosted transcription via the A- and B-box in the tRNA.

2.6 Cotransfection of Cas9 Protein Loaded With Different gRNAs

Recently, a DNA-free approach for CRISPR/Cas9 genome editing was developed in various organisms including plants.[46–49] Purified Cas9 endonuclease is incubated with in vitro transcribed gRNA to form preassembled Cas9-gRNA ribonucleoproteins called RNA-guided endonuclease ribonucleoproteins (RGEN RNPs, Fig. 1F). These RGEN RNPs have several advantages because they do not require a plasmid vector or other forms of DNA for expression and are quickly degraded after delivery into cells and editing the genome.[46–48] This feature is especially valuable for crop trait improvement because genome-editing RNPs will produce a mutation without the chance of foreign DNA integration. The *Agrobacterium*-mediated transformation leads to transfer DNA (T-DNA) integration from which Cas9 and gRNA are expressed. Particle bombardment of expression vectors also leads to random integration of the plasmid or DNA fragments into the genome. While these foreign DNA integrations could be removed by backcrossing or selfing, this is not feasible if generation times are long or if the crop is propagated vegetatively.

RGEN RNPs-mediated multiplex editing can be performed by simply mixing RNPs that are loaded with different gRNAs. Kim *et al.* showed that they can delete chromosomal fragments in the sizes of 10, 30, or even 100 kb in human K562 cell by supplying RGEN RNPs loaded with two distinct gRNAs.[47] Similarly, multiplexing capability with two gRNAs for RGEN RNPs was shown in plants by a deletion of a short 223 bp sequence in *Arabidopsis*.[48] Woo *et al.* tested RGEN RNPs in protoplasts of *Arabidopsis*, tobacco, and rice with mutation frequencies of 16%–19%, 17%–44%, and 8.4%–23%, respectively, depending on the method of indel detection.[48] The same study also reported successful production of genome-edited lettuce plants (*Lactuca sativa*) after introduction of RGEN RNPs into lettuce protoplasts and then regeneration of microcalli with an overall mutation frequency of 46%.[48]

While mutated lettuce could be readily regenerated from protoplasts, this method is not feasible for many other important plant and crop species that have inefficient plant regeneration from protoplasts. Two recent studies in maize and wheat, however, circumvented this problem by loading RGEN RNPs to tungsten particles. These particles can be used for particle bombardment of tissue and therefore allow, for example, to perform DNA-free editing on totipotent plant tissue-like embryogenic callus. When maize embryos were cobombarded with RGEN RNPs, transformation-enhancing "helper genes," and a selectable and visible marker (MoPAT-DsRED), 47% of the obtained herbicide resistant plants were successfully mutated.[50] They also achieved homology directed repair (HDR) of the maize *ALS2* gene by supplying an additional 127 nt single-stranded DNA oligo as repair template. The authors of the study in wheat did not select with a marker and were still able to achieve a mutation efficiency of 4.4% in 640 screened wheat plantlets.[51] These promising results demonstrate that genome-edited mutants can be obtained via DNA-free RGEN RNPs with high efficiencies. Mutants resulting from this method are likely to bypass current regulatory hurdles because they do not carry any foreign gene fragments but only small deletions or insertions in the target region. However, whether RGEN RNPs loaded to tungsten particles can efficiently mediate multiplex editing still needs to be investigated.

3. BROAD APPLICATIONS OF MULTIPLEX GENOME EDITING IN PLANTS

The development of CIRSPR/Cas9 genome editing and its ability to perform multiplex editing with high efficiencies has opened doors for a plethora of exciting applications in plant biology and crop breeding. For example, the ability to mutate several genes at once allows for analysis of gene families with members that may have redundant functions. Large fragments of genes or chromosomes can be deleted when two gRNAs are simultaneously introduced. Multiplexed CRISPR/Cas9 even allows for more robust gene activation or repression because several activators or repressors can be recruited to the target gene. Fig. 2 provides a stylistic depiction of

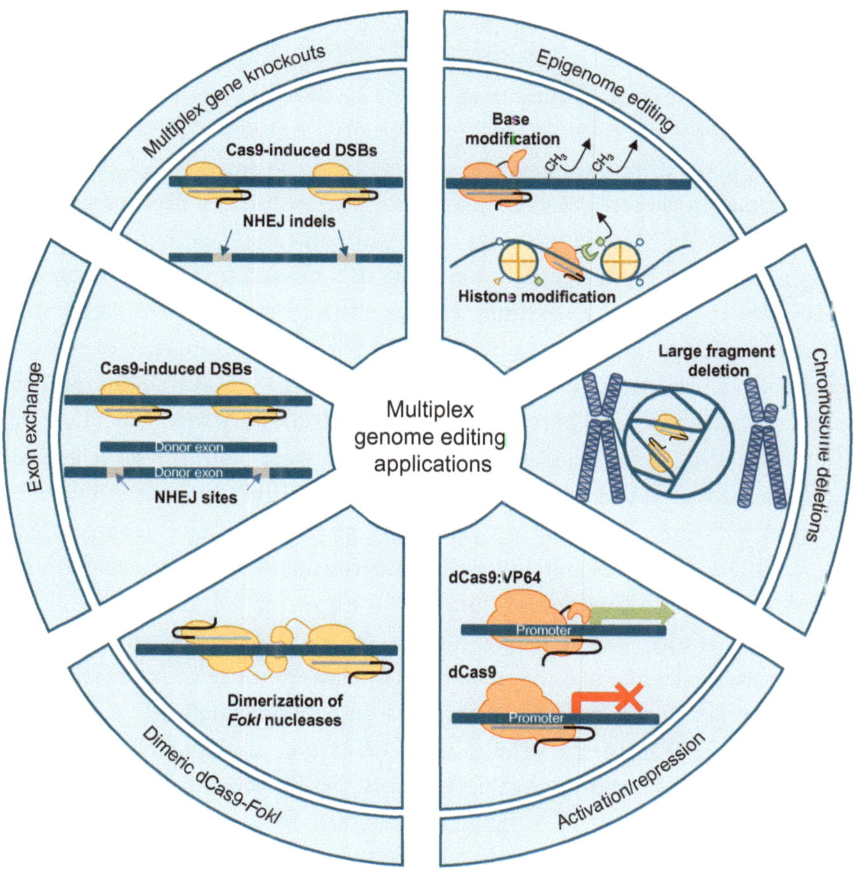

Fig. 2 Various applications of multiplex genome editing.

these applications while the following section provides an overview about the current and future applications that are enabled by CRISPR/Cas9's multiplexing capabilities.

3.1 Multigene Knockout via NHEJ

The multiplex genome-editing application that has been mostly exploited in plants is the creation of plants with several gene knockouts. *Arabidopsis* multigene knockout plants were previously created, for example, by first identifying reliable mutant lines from either T-DNA insertion libraries or chemical mutagenesis screens. Then researchers needed to cross the mutant lines in a time-consuming process to obtain homozygous double, triple, or even higher-tuple mutants. CRISPR/Cas9's multiplexing capabilities allow for creation of higher-tuple mutants with only one transformation event.

Proteins of the *PYL/PYR* gene family are involved in abscisic acid (ABA) perception and consist of 14 members.[52,53] Researchers needed to create quadruple and sextuple mutants of the *PYL/PYR* gene family member in order to achieve high ABA insensitivity mutants. A study using CRISPR/Cas9 multiplex genome editing tried to create a *PYL/PYR* sextuple mutant using one construct harboring expression cassettes for six independent gRNAs. Homozygous sextuple mutants could be obtained in the T3 generation.[12] A comparison between the crossed sextuple mutant and the CRISPR-generated sextuple mutant showed that both mutants were resistant to ABA-mediated inhibition of root growth, but the previously described crossed mutant showed a higher effect.[12,53] While the authors found slight differences between the two sextuple mutants, only five out of six mutations in the crossed sextuple mutant were from the same genetic background.[12,53] It is unclear if this contributed to the observed differences.

Another study aimed to analyze genes of the rice *MAP kinase* (*MKP*) genes. MPKs are important signal transducers in plants for biotic and abiotic stresses as well as development and phytohormonal signals. Most of the knowledge about plant MPK signaling stems from *Arabidopsis* mutants, whereas studies in rice mainly relied on RNA interference or overexpression lines. Several constructs were created to target combinations of four rice *MPKs* with PTGs to create a library of single, double and quadruple mutants.[44] Double and quadruple mutants were obtained even in the T0 generation with up to 45% biallelic mutation frequency for the double mutant and 86% for the quadruple mutant. Functional analysis of the obtained *MPK* mutants revealed essentiality in rice for two of the four targeted genes, which was in contrast to *Arabidopsis* orthologues, where homozygous *MPK* knockout lines remain viable.[44]

The aforementioned studies showed that multiple order mutants can be readily obtained by CRISPR/Cas9 multiplex genome editing. The ability to produce such mutants promises to drastically speed up genetic analysis in basic research. Members of gene families sometimes have redundant or overlapping functions, and knockouts of single genes often give rise to weak phenotypes. Simultaneous knockout of several gene family members may lead to the discovery of new and more drastic phenotypes. *Arabidopsis* is a popular model system for plant genetics because of the availability of T-DNA mutant libraries. CRISPR/Cas9 genome editing, however, is applicable for almost any crop as long as Cas9 and the gRNA can be delivered as vectors or protein complexes, which greatly expands the reach of genetic analysis to many different species. It is expected that many new gene functions in plants other than model species will be discovered in the future. This is especially important for many crop species that lack mutant resources until now.

3.2 Deletion of Chromosomal or Gene Fragments

The expression of two gRNAs that target regions on the same chromosome or gene enable the deletion of chromosomal fragments. Whole genes or parts of them, regulatory elements, gene clusters, or large chromosomal fragment can be deleted from the genome. Large chromosomal deletion of up to 245 kb was reported with this strategy.[54] In general, the deletion efficiency will decline with expanded distance between the two target sites.[9] For example, a 358-bp fragment could be deleted with 45% efficiency in rice protoplasts, whereas a 507-bp deletion could only be achieved at 26% efficiency.[9] A recent study deleted a regulatory element in an intron from the *Arabidopsis* flower development gene *AGAMOUS* (*AG*). They successfully obtained a 450-bp deletion in the second intron of the gene in 25% of the transformants.[19] Deletion of the regulatory elements in this gene reduced expression of *AG* to 60% of the wild-type level and produced a partial *ag* phenotype.[19]

Deletions of bigger fragments can be more useful compared to the usually short insertions or deletions produced when targeting with only one gRNA. As shown with *AG*, the fragment deletion allows the functional study of regulatory elements as these can be cut out more precisely. Targeting with two gRNAs also provide a more reliable way to achieve gene knockout because whole gene can be deleted. Deletions of a few hundred base pairs can be easily screened in PCR assays because the precise size difference between the mutant and wild-type alleles. In addition to knockout of protein coding genes, fragment deletion is also valuable for studying small RNAs or long noncoding RNAs. Since these are acting as RNAs and do not get translated, a deletion is necessary to effectively knockout their functions.

3.3 Exon Exchange via NHEJ

Targeting a gene with CRISPR/Cas9 alone will usually lead to a gene knockout, because the main pathway to repair the double strand break in higher organisms is the nonhomologous end-joining (NHEJ) pathway. This DNA repair pathway produces small insertion or deletions at the target sites because of its error-prone nature. When researchers want to introduce precise changes to the genome, they usually try to exploit a different DNA repair pathway called HDR. This pathway can achieve exact change of nucleotides if a repair template is provided. However, the HDR pathway in plants is very inefficient and introducing precise changes that, for example, alter the protein sequence are still hard to achieve and require screening of high number of events.[1]

A recent study described an interesting application of NHEJ-based multiplex genome editing for exon replacement that can bypass the need of the inefficient HDR pathway. Li *et al.* replaced the second exon of the rice *ESPS* gene with a version that contains two amino acid substitutions (T102I and P106S) that leads to a herbicide resistance phenotype.[55] They targeted the exon by inducing double strand breaks in the introns flanking the target exon. Simultaneously, a replacement exon was cut out from the donor plasmid. This strategy leads to successful replacement of the targeted exon with the new version in 2% of the tested plants.[55]

3.4 Gene Activation and Repression

CRISPR/Cas9 can be used for gene activation or gene repression with some modifications. Gene activation is achieved if a catalytically dead Cas9 version (dCas9) without nuclease activity is fused to an activation domain such as VP64.[21,56,57] Gene repression with CRISPR/Cas9 follows a similar strategy by targeting a gene with either dCas9 alone or dCas9 coupled to repressors to reduce expression.[11,58] Recruitment of these Cas9-based activators or repressors with a single gRNA is insufficient for robust activation or repression. Multiplex genome editing enhances gene activation and repression because several of these Cas9-based transcription factors can be recruited to the target site for an additive effect. Expression of multiple effectors could also allow the fine-tuning of gene expression in plants to enable metabolic engineering for altered or enhanced production of metabolites.

While most of these approaches were developed in mammalian cells, one study tested CRISPR/Cas9-mediated gene activation and repression in plants. Piatek *et al.* fused dCas9 with the activation domain of EDLL and TAL effectors for gene activation and fused dCas9 with the SDRX repression domain for gene repression.[59] The different dCas9-based transcription factors were tested using agroinfiltration of *N. benthamiana* leaves. They showed successful activation and suppression of the endogenous *PDS* gene and an substantial synergistic effect when codelivering multiple gRNAs, further demonstrating that multiplex genome editing enables more robust manipulation of transcript expression.[59]

3.5 Cas9-Dimers to Reduce Off-Targets

Since CRISPR/Cas9 finds its target via RNA/DNA interaction between the gRNA spacer and the target DNA, CRISPR/Cas9 can have off-target activity at places in the genome with sufficient sequence similarity.[60–63] Previous studies identified three main types of off-targets for the most widely used SpCas9 from *S. pyogenes*. SpCas9 can cut sequences if they are similar enough to the gRNA spacer on-target region and near a 5′-NGG-3′ PAM.[61,63,64] The second type is caused by sequences that have insertions or deletions when compared to the on-target sequence.[65] The third type of off-targets that can be cleaved by Cas9 are similar sequences near a 5′-NAG-3′ PAM.[61,63,66]

Multiplex genome editing, however, can alleviate some of the off-target concerns by transforming the CRISPR/Cas9 system into a two-component system. This can be achieved by fusing dCas9 with a dimerization-dependent *Fok*I domain.[28] Cleavage of the target site is only possible when both *Fok*I domains are in close proximity to each other. Another approach uses Cas9-nickases that also require two of the proteins to be in close proximity to achieve efficient double strand breaks.[67] Because of these restraints, both modifications need simultaneous expression of two gRNAs to achieve efficient genome editing. The requirement of two gRNAs increases specificity because the DNA/RNA binding requirement is lifted up from 20 to 40 nt, and it is less likely that both 20 nt spacers will have an additional off-target in such close proximity elsewhere in the genome.

A study in plants confirmed that paired Cas9-nickases can efficiently mutate the *Arabidopsis* genome and mutations are inherited into future generations.[68] *Arabidopsis RETL1* locus was targeted with a Cas9-nickase guided by two gRNAs and resulting T1 transgenic plants were analyzed by amplicon deep sequencing. While the study did not look at off-target mutagenesis, it at least showed that dCas9-nickases can efficiently mutate genes in plants. Based on a subset of 250 reads, the mutation frequency was 42.8% with a wide range of mutations types.[69] Another study in rice provided a more detailed comparison between wild-type Cas9 nuclease and paired Cas9-nickases. They targeted two sites in close proximity with either paired nucleases that induce two double strand breaks or paired nickases that induce single strand breaks and evaluated on-target as well as off-target activities.[70] Mutation frequencies in regenerated plantlets transformed with paired wild-type Cas9 was 62.5%–87.5%, but it reduced to 53.1%–59.3% when nickases were used instead, showing that paired Cas9-nickases have lower on-target mutation frequency compared to wild-type Cas9 nucleases.[70] On the other hand, they showed that off-target mutations were drastically reduced to an undetectable level when using Cas9-nickases even if one of the paired Cas9-nickases had a perfect match in a different genomic region.[70] Cas9-nickases are therefore a reliable tool to avoid off-target mutations in plants and they are empowered by the multiplexing capabilities of the CRISPR/Cas9 system. This might be especially useful for genome-editing tasks in plants that lack high quality genome sequence resources so far.

3.6 Epigenome Editing

The last application for multiplex genome editing with CRISPR/Cas9 that we will discuss here is epigenome editing. Similar to some of the previous examples, dCas9 is coupled to other domains or proteins to enable targeted enzyme activity to change DNA methylation or histone modifications. Changes in DNA methylation states are promising for plant trait development because they can be inherited to future generations without changing the DNA itself. Manipulation of DNA methylation presents therefore a completely new way of mutation free trait engineering that can be revolutionized by multiplexed CRISPR/Cas9 application.

For DNA and histone modifications, enzymes can be used to add or remove epigenetic marks, respectively. Recent studies in mammalian organisms showed that dCas9 can be coupled to these enzymes for nucleotide modifications. Specifically, DNA methyltransferase 3a coupled to dCas9 (dCas9-Dnmt3) was shown to induce methylation in targeted promoter regions resulting in gene silencing.[71] Conversely, a demethylation enzyme, 10-11 translocation dioxygenase 1, has been fused to dCas9 (dCas9-Tet1) for inducing demethylation within a promoter region to achieve target gene activation.[71]

Histone modifications have been achieved by fusing histone modifying enzymes to dCas9 to achieve sequence-specific epigenetic modifications.[71] The core active domains of the acetyltransferase p300 was fused to dCas9 for creating $dCas9^{p300Core}$ to increase H3K27 acetylation.[72] LSD1, a histone demethylase, was fused to dCas9 resulting in a decreased amount of H3K4 methylation modifications.[72] While these are only a few examples, new tools for epigenome editing are being developed constantly to further expand the possibilities to change gene expression states. These valuable tools are expected to facilitate basic research and crop improvement, especially when combined with multiplexed editing to achieve more robust changes or changes on multiple targets.

4. CONCLUSION

Diverse strategies have been developed for expression of multiple gRNAs and simultaneous editing of multiple targets in plant genome. Multiplex genome editing has already shown broad applications in plant biology and is expected to greatly facilitate crop genome engineering and precision breeding. DNA-free genome editing via RGEN RNPs and other transgene-free approaches are especially promising for commercialization of gene-edited crops because the resulting plants would not contain foreign DNA. The future of multiplex genome editing is likely not limited by multiplex capability itself but rather the invention of various new ways to use the current and future CRISPR/Cas editing tools. The innovation and application of new methods and tools for multiplex genome editing are expect to have profound impacts on basic studies in plant biology and applied research for crop improvement.

ACKNOWLEDGMENTS

This work was supported by Monsanto's Beachell-Borlaug International Scholars Program and Penn State's J. Lloyd Huck Dissertation Research Grant to B.M. This work was also supported by Agriculture and Food Research Initiative Competitive Grant No. 2013-68004-20378 from the USDA National Institute of Food and Agriculture.

REFERENCES

1. Voytas DF. Plant genome engineering with sequence-specific nucleases. *Annu Rev Plant Biol.* 2013;64:327–350. http://dx.doi.org/10.1146/annurev-arplant-042811-105552.
2. Kim YG, Cha J, Chandrasegaran S. Hybrid restriction enzymes: zinc finger fusions to Fok I cleavage domain. *Proc Natl Acad Sci USA.* 1996;93(3):1156–1160. http://dx.doi.org/10.1073/pnas.93.3.1156.
3. Ramirez CL, Foley JE, D a W, et al. Unexpected failure rates for modular assembly of engineered zinc fingers. *Nat Methods.* 2008;5(5):374–375. http://dx.doi.org/10.1038/nmeth0508-374.
4. Moscou MJ, Bogdanove AJ. A simple cipher governs DNA recognition by TAL effectors. *Science.* 2009;326(5959):1501. http://dx.doi.org/10.1126/science.1178817.
5. Boch J, Scholze H, Schornack S, et al. Breaking the code of DNA binding specificity of TAL-type III effectors. *Science.* 2009;326(5959):1509–1512. http://dx.doi.org/10.1126/science.1178811.
6. Jinek M, Chylinski K, Fonfara I, Hauer M, Doudna JA, Charpentier E. A programmable dual-RNA-guided DNA endonuclease in adaptive bacterial immunity. *Science.* 2012;337(6096):816–821. http://dx.doi.org/10.1126/science.1225829.
7. Mali P, Yang L, Esvelt KM, et al. RNA-guided human genome engineering via Cas9. *Science.* 2013;339:823–827.
8. Cong L, Ran FA, Cox D, et al. Multiplex genome engineering using CRISPR/Cas systems. *Science.* 2013;339:819–824.
9. Xie K, Minkenberg B, Yang Y. Boosting CRISPR/Cas9 multiplex editing capability with the endogenous tRNA-processing system. *Proc Natl Acad Sci USA.* 2015;112(11):3570–3575. http://dx.doi.org/10.1073/pnas.1420294112.
10. Xing H-L, Dong L, Wang Z-P, et al. A CRISPR/Cas9 toolkit for multiplex genome editing in plants. *BMC Plant Biol.* 2014;14(1):327. http://dx.doi.org/10.1186/s12870-014-0327-y.
11. Vazquez-Vilar M, Bernabé-Orts JM, Fernandez-Del-Carmen A, et al. A modular toolbox for gRNA-Cas9 genome engineering in plants based on the GoldenBraid standard. *Plant Methods.* 2016;12:10. http://dx.doi.org/10.1186/s13007-016-0101-2.
12. Zhang Z, Mao Y, Ha S, Liu W, Botella JR, Zhu JK. A multiplex CRISPR/Cas9 platform for fast and efficient editing of multiple genes in Arabidopsis. *Plant Cell Rep.* 2016;35(7):1519–1533. http://dx.doi.org/10.1007/s00299-015-1900-z.
13. Ma X, Zhang Q, Zhu Q, et al. A robust CRISPR/Cas9 system for convenient, high-efficiency multiplex genome editing in monocot and dicot plants. *Mol Plant.* 2015;8(8):1274–1284. http://dx.doi.org/10.1016/j.molp.2015.04.007.
14. Wang C, Shen L, Fu Y, Yan C, Wang K. A simple CRISPR/Cas9 system for multiplex genome editing in rice. *J Genet Genomics.* 2015;42(12):703–706. http://dx.doi.org/10.1016/j.jgg.2015.09.011.
15. Char SN, Neelakandan AK, Nahampun H, et al. An Agrobacterium-delivered CRISPR/Cas9 system for high-frequency targeted mutagenesis in maize. *Plant Biotechnol J.* 2017;15:257–268. http://dx.doi.org/10.1111/pbi.12611.
16. Zhao Y, Zhang C, Liu W, et al. An alternative strategy for targeted gene replacement in plants using a dual-sgRNA/Cas9 design. *Sci Rep.* 2016;6:23890. http://dx.doi.org/10.1038/srep23890.
17. Cao J, Wu L, Zhang SM, et al. An easy and efficient inducible CRISPR/Cas9 platform with improved specificity for multiple gene targeting. *Nucleic Acids Res.* 2016;44(19):1–10. http://dx.doi.org/10.1093/nar/gkw660.
18. Tian S, Jiang L, Gao Q, et al. Efficient CRISPR/Cas9-based gene knockout in watermelon. *Plant Cell Rep.* 2016;36(3):399–406. http://dx.doi.org/10.1007/s00299-016-2089-5.

19. Yan W, Chen D, Kaufmann K. Efficient multiplex mutagenesis by RNA-guided Cas9 and its use in the characterization of regulatory elements in the AGAMOUS gene. *Plant Methods*. 2016;12(1):23. http://dx.doi.org/10.1186/s13007-016-0125-7.
20. Ordon J, Gantner J, Kemna J, et al. Generation of chromosomal deletions in dicotyledonous plants employing a user-friendly genome editing toolkit. *Plant J*. 2017;89: 155–168. http://dx.doi.org/10.1111/tpj.13319.
21. Lowder LG, Zhang D, Baltes NJ, et al. A CRISPR/Cas9 toolbox for multiplexed plant genome editing and transcriptional regulation. *Plant Physiol*. 2015;169:971–985. http:// dx.doi.org/10.1104/pp.15.00636.
22. Li J-F, Norville JE, Aach J, et al. Multiplex and homologous recombination-mediated genome editing in Arabidopsis and Nicotiana bethamiana using guide RNA and Cas9. *Nat Biotechnol*. 2013;31(8):688–691. http://dx.doi.org/10.1038/nbt.2650.
23. Peterson BA, Haak DC, Nishimura MT, et al. Genome-wide assessment of efficiency and specificity in CRISPR/Cas9 mediated multiple site targeting in *Arabidopsis*. *PLoS One*. 2016;11(9): e0162169. http://dx.doi.org/10.1371/journal.pone.0162169.
24. Andersson M, Turesson H, Nicolia A, Fält AS, Samuelsson M, Hofvander P. Efficient targeted multiallelic mutagenesis in tetraploid potato (Solanum tuberosum) by transient CRISPR-Cas9 expression in protoplasts. *Plant Cell Rep*. 2017;36(1):117–128. http://dx. doi.org/10.1007/s00299-016-2062-3.
25. Nomura T, Sakurai T, Osakabe Y, Osakabe K, Sakakibara H. Efficient and heritable targeted mutagenesis in mosses using the CRISPR/Cas9 system. *Plant Cell Physiol*. 2016;57:pcw173. http://dx.doi.org/10.1093/pcp/pcw173.
26. Haurwitz RE, Jinek M, Wiedenheft B, Zhou K, Doudna JA. Sequence- and structure-specific RNA processing by a CRISPR endonuclease. *Science*. 2010;329:1355–1358. http://dx.doi.org/10.1126/science.1192272.
27. Haurwitz RE, Sternberg SH, Doudna JA. Csy4 relies on an unusual catalytic dyad to position and cleave CRISPR RNA. *EMBO J*. 2012;31(12):2824–2832. http://dx.doi. org/10.1038/emboj.2012.107.
28. Tsai SQ, Wyvekens N, Khayter C, et al. Dimeric CRISPR RNA-guided FokI nucleases for highly specific genome editing. *Nat Biotechnol*. 2014;32(6):569–576. http://dx.doi. org/10.1038/nbt.2908.
29. Zetsche B, Heidenreich M, Mohanraju P, et al. Multiplex gene editing by CRISPR–Cpf1 using a single crRNA array. *Nat Biotechnol*. 2017;35:31–34. http://dx.doi.org/ 10.1038/nbt.3737.
30. Zetsche B, Gootenberg JS, Abudayyeh OO, et al. Cpf1 is a single RNA-guided endonuclease of a Class 2 CRISPR-Cas system. *Cell*. 2015;163(3):759–771. http://dx.doi. org/10.1016/j.cell.2015.09.038.
31. Xu R, Qin R, Li H, et al. Generation of targeted mutant rice using a CRISPR-Cpf1 system. *Plant Biotechnol J*. 2017;15:713–717. http://dx.doi.org/10.1111/pbi.12669.
32. Endo A, Masafumi M, Kaya H, Toki S. Efficient targeted mutagenesis of rice and tobacco genomes using Cpf1 from Francisella novicida. *Sci Rep*. 2016;6:38169. http://dx.doi.org/10.1038/srep38169.
33. Tang X, Lowder LG, Zhang T, et al. A CRISPR–Cpf1 system for efficient genome editing and transcriptional repression in plants. *Nat Plants*. 2017;3:17018. http://dx. doi.org/10.1038/nplants.2017.18.
34. Gao Y, Zhao Y. Self-processing of ribozyme-flanked RNAs into guide RNAs in vitro and in vivo for CRISPR-mediated genome editing. *J Integr Plant Biol*. 2014;56(4):343–349. http://dx.doi.org/10.1111/jipb.12152.
35. Lee RTH, Ng ASM, Ingham PW. Ribozyme mediated gRNA generation for in vitro and in vivo CRISPR/Cas9 mutagenesis. *PLoS One*. 2016;11(11):1–12. http://dx doi. org/10.1371/journal.pone.0166020.

36. Zhang WW, Matlashewski G. CRISPR-Cas9-mediated genome editing in Leishmania donovani. *MBio*. 2015;6(4):1–14. http://dx.doi.org/10.1128/mBio.00861-15.
37. Xu L, Zhao L, Gao Y, Xu J, Han R. Empower multiplex cell and tissue-specific CRISPR-mediated gene manipulation with self-cleaving ribozymes and tRNA. *Nucleic Acids Res*. 2016;45(5):gkw1048. http://dx.doi.org/10.1093/nar/gkw1048.
38. Wang W, Akhunova A, Chao S, Akhunov E. Optimizing multiplex CRISPR/Cas9-based genome editing for wheat. *bioRxiv*. 2016. https://dx.doi.org/10.1101/051342.
39. Qi W, Zhu T, Tian Z, Li C, Zhang W, Song R. High-efficiency CRISPR/Cas9 multiplex gene editing using the glycine tRNA-processing system-based strategy in maize. *BMC Biotechnol*. 2016;16(1):58. http://dx.doi.org/10.1186/s12896-016-0289-2.
40. Port F, Bullock SL. Augmenting CRISPR applications in Drosophila with tRNA-flanked sgRNAs. *Nat Methods*. 2016;4:1–5. http://dx.doi.org/10.1038/nmeth.3972.
41. Dong F, Xie K, Chen Y, Yang Y, Mao Y. Polycistronic tRNA and CRISPR guide-RNA enables highly efficient multiplexed genome engineering in human cells. *Biochem Biophys Res Commun*. 2017;482(4):889–895. http://dx.doi.org/10.1016/j.bbrc.2016.11.129.
42. Schwartz CM, Hussain MS, Blenner M, Wheeldon I. Synthetic RNA polymerase III promoters facilitate high-efficiency CRISPR-Cas9-mediated genome editing in Yarrowia lipolytica. *ACS Synth Biol*. 2016;5(4):356–359. http://dx.doi.org/10.1021/acssynbio.5b00162.
43. White RJ. Transcription by RNA polymerase III: more complex than we thought. *Nat Rev Genet*. 2011;12(7):459–463. http://dx.doi.org/10.1038/nrg3001.
44. Minkenberg B, Xie K, Yang Y. Discovery of rice essential genes by characterizing CRISPR-edited mutation of closely related rice MAP kinase genes. *Plant J*. 2017;89:636–648. http://dx.doi.org/10.1111/tpj.13399.
45. Yan Q, Xu K, Xing J, et al. Multiplex CRISPR/Cas9-based genome engineering enhanced by Drosha-mediated sgRNA-shRNA structure. *Sci Rep*. 2016;6:38970. http://dx.doi.org/10.1038/srep3870.
46. Cho SW, Lee J, Carroll D, Kim JS, Lee J. Heritable gene knockout in Caenorhabditis elegans by direct injection of Cas9-sgRNA ribonucleoproteins. *Genetics*. 2013;195(3):1177–1180. http://dx.doi.org/10.1534/genetics.113.155853.
47. Kim S, Kim D, Cho SW, Kim J, Kim J-S. Highly efficient RNA-guide genome editing in human cells via delivery of purified Cas9 ribonucleoproteins. *Genome Res*. 2014;24:1012–1019. http://dx.doi.org/10.1101/gr.171322.113.
48. Woo JW, Kim J, Kwon SI, et al. DNA-free genome editing in plants with preassembled CRISPR-Cas9 ribonucleoproteins. *Nat Biotechnol*. 2015;33(11):1162–1164. http://dx.doi.org/10.1038/nbt.3389.
49. Baek K, Kim DH, Jeong J, et al. DNA-free two-gene knockout in Chlamydomonas reinhardtii via CRISPR-Cas9 ribonucleoproteins. *Sci Rep*. 2016;6:30620. http://dx.doi.org/10.1038/srep30620.
50. Svitashev S, Schwartz C, Lenderts B, Young JK, Mark Cigan A. Genome editing in maize directed by CRISPR-Cas9 ribonucleoprotein complexes. *Nat Commun*. 2016;7:13274. http://dx.doi.org/10.1038/ncomms13274.
51. Liang Z, Chen K, Li T, et al. Efficient DNA-free genome editing of bread wheat using CRISPR/Cas9 ribonucleoprotein complexes. *Nat Commun*. 2017;8:14261. http://dx.doi.org/10.1038/ncomms14261.
52. Park S-Y, Fung P, Nishimura N, et al. Abscisic acid inhibits type 2C protein phosphatases via the PYR/PYL family of START proteins. *Science*. 2009;324(5930):1068–1071. http://dx.doi.org/10.1126/science.1173041.

53. Gonzalez-Guzman M, Pizzio GA, Antoni R, et al. Arabidopsis PYR/PYL/RCAR receptors play a major role in quantitative regulation of stomatal aperture and transcriptional response to abscisic acid. *Plant Cell.* 2012;24(6):2483–2496. http://dx.doi.org/10.1105/tpc.112.098574.
54. Zhou H, Liu B, Weeks DP, Spalding MH, Yang B. Large chromosomal deletions and heritable small genetic changes induced by CRISPR/Cas9 in rice. *Nucleic Acids Res.* 2014;42(17):10903–10914. http://dx.doi.org/10.1093/nar/gku806.
55. Li J, Meng X, Zong Y, et al. Gene replacements and insertions in rice by intron targeting using CRISPR-Cas9. *Nat Plants.* 2016;2:16139. http://dx.doi.org/10.1038/nplants.2016.139.
56. Perez-Pinera P, Kocak DD, Vockley CM, et al. RNA-guided gene activation by CRISPR-Cas9-based transcription factors. *Nat Methods.* 2013;10(10):973–976. http://dx.doi.org/10.1038/nmeth.2600.
57. Mali P, Aach J, Stranges PB, et al. CAS9 transcriptional activators for target specificity screening and paired nickases for cooperative genome engineering. *Nat Biotechnol.* 2013;31(9):833–838. http://dx.doi.org/10.1038/nbt.2675.
58. Gilbert LA, Larson MH, Morsut L, et al. CRISPR-mediated modular RNA-guided regulation of transcription in eukaryotes. *Cell.* 2013;154(2):442–451. http://dx.doi.org/10.1016/j.cell.2013.06.044.
59. Piatek A, Ali Z, Baazim H, et al. RNA-guided transcriptional regulation in planta via synthetic dCas9-based transcription factors. *Plant Biotechnol J.* 2015;13(4):578–589. http://dx.doi.org/10.1111/pbi.12284.
60. Cradick TJ, Fine EJ, Antico CJ, Bao G. CRISPR/Cas9 systems targeting β-globin and CCR5 genes have substantial off-target activity. *Nucleic Acids Res.* 2013;41(20):9584–9592. http://dx.doi.org/10.1093/nar/gkt714.
61. Fu Y, Foden JA, Khayter C, et al. High-frequency off-target mutagenesis induced by CRISPR-Cas nucleases in human cells. *Nat Biotechnol.* 2013;31(9):822–826. http://dx.doi.org/10.1038/nbt.2623.
62. Xie K, Yang Y. RNA-guided genome editing in plants using a CRISPR-Cas system. *Mol Plant.* 2013;6:1975–1983. http://dx.doi.org/10.1093/mp/sst119.
63. Tsai SQ, Zheng Z, Nguyen NT, et al. GUIDE-seq enables genome-wide profiling of off-target cleavage by CRISPR-Cas nucleases. *Nat Biotechnol.* 2015;33(2):187–197. http://dx.doi.org/10.1038/nbt.3117.
64. Pattanayak V, Lin S, Guilinger JP, Ma E, Doudna JA, Liu DR. High-throughput profiling of off-target DNA cleavage reveals RNA-programmed Cas9 nuclease specificity. *Nat Biotechnol.* 2013;31(9):839–843. http://dx.doi.org/10.1038/nbt.2673.
65. Lin Y, Cradick TJ, Brown MT, et al. CRISPR/Cas9 systems have off-target activity with insertions or deletions between target DNA and guide RNA sequences. *Nucleic Acids Res.* 2014;42(11):7473–7485. http://dx.doi.org/10.1093/nar/gku402.
66. Hsu PD, Scott DA, Weinstein JA, et al. DNA targeting specificity of RNA-guided Cas9 nucleases. *Nat Biotechnol.* 2013;31(9):827–832. http://dx.doi.org/10.1038/nbt.2647.
67. Shen B, Zhang W, Zhang J, et al. Efficient genome modification by CRISPR-Cas9 nickase with minimal off-target effects. *Nat Methods.* 2014;11(4):399–402. http://dx.doi.org/10.1038/nmeth.2857.
68. Schiml S, Fauser F, Puchta H. The CRISPR/Cas system can be used as nuclease for in planta gene targeting and as paired nickases for directed mutagenesis in Arabidopsis resulting in heritable progeny. *Plant J.* 2014;80(6):1139–1150. http://dx.doi.org/10.1111/tpj.12704.
69. Fauser F, Schiml S, Puchta H. Both CRISPR/Cas-based nucleases and nickases can be used efficiently for genome engineering in Arabidopsis thaliana. *Plant J.* 2014;79(2):348–359. http://dx.doi.org/10.1111/tpj.12554.

70. Mikami M, Toki S, Endo M. Precision targeted mutagenesis via Cas9 paired nickases in rice. *Plant Cell Physiol.* 2016;57(5):1058–1068. http://dx.doi.org/10.1093/pcp/pcw049.
71. Liu XS, Wu H, Ji X, et al. Editing DNA methylation in the mammalian genome. *Cell.* 2016;167(1):233–247.e17. http://dx.doi.org/10.1016/j.cell.2016.08.056.
72. Thakore PI, Black JB, Hilton IB, Gersbach CA. Editing the epigenome: technologies for programmable transcription and epigenetic modulation. *Nat Methods.* 2015;13:127–137. http://dx.doi.org/10.1038/nmeth.3733.

CHAPTER EIGHT

CRISPR/Cas9-Based Genome Editing in Plants

Yaling Zhang*,†,[1], Xingliang Ma‡,[1], Xianrong Xie*,†,
Yao-Guang Liu*,†,[2]
*State Key Laboratory for Conservation and Utilization of Subtropical Agro-Bioresources, Guangzhou, China
†College of Life Sciences, South China Agricultural University, Guangzhou, China
‡Global Institute for Food Security, University of Saskatchewan, Saskatoon, SK, Canada
[2]Corresponding author: e-mail address: ygliu@scau.edu.cn

Contents

[1] These authors contributed equally to this chapter.

Progress in Molecular Biology and Translational Science, Volume 149
ISSN 1877-1173
http://dx.doi.org/10.1016/bs.pmbts.2017.03.008

Abstract

Recently, genome editing technologies have shown great potential in plants. The newly developed clustered regularly interspaced short palindromic repeats (CRISPR)/Cas9 system is a new generation of genome editing tool rapidly replacing the earlier zinc finger nucleases and transcription activator-like effector nucleases systems. Indeed, due to its advantages of simplicity and high efficiency, the CRISPR/Cas9-based genome editing system is becoming a powerful tool in plant science research. Here, we introduce the technical features of the plant CRISPR/Cas9-based genome editing system and its applications in plant functional genomics studies and genetic improvement.

1. INTRODUCTION

Genetic mutants play very important roles in gene functional research of plants and crop improvement. In life science research, many important biological mechanisms have been revealed using mutants. Traditionally, random mutants are obtained from natural, physical, chemical, or biological (e.g., T-DNA/transposon insertion) mutagenesis, followed by large-scale screening of mutations from tens of thousands of genes in model plants, such as *Arabidopsis*[1] and rice.[2,3] However, many undesirable mutations can be caused during traditional random mutagenesis process, and large-scale screening of intended mutations is time-consuming and costly.[4]

Genome editing technologies refer to those having the ability to perform knockout, chromosomal recombination and site-directed insertion/substitution at specific sites of genes and chromosomal regions. In brief, genome editing technologies produce double-stranded breaks (DSBs) in DNA of specific chromosomal sites through sequence-specific nucleases (SSNs), and the DSBs are subsequently repaired either by the error-prone nonhomologous end joining (NHEJ) pathway causing nucleotide insertions or deletions or by the homology-directed repair (HDR) pathway that directs DNA sequence substitutions in the sites.[5] If an error during DNA repair occurs in the coding region of the gene, it may cause codon mutations or frameshift mutations of the gene. Compared to random mutagenesis, genome editing technologies can introduce mutations at specific chromosomal sites, thus have many advantages in functional genomics and molecular breeding. Indeed, genome fixed-point editing techniques not only have a greater chance to cause mutations but also are more convenient and more powerful than random mutagenesis. Moreover, simple genetic segregation enables production of transgene-free plants during crop breeding programs.[6,7]

1.1 Types of Genome Editing Technologies

So far, genome editing technologies comprise three primary SSN systems. These are the zinc finger nucleases (ZFNs), the transcription activator-like effector nucleases (TALENs), and the clustered regularly interspaced short palindromic repeats/CRISPR-associated protein 9 (CRISPR/Cas9) system. The ZFNs were the first generation of SSNs used to edit plant genomes.[8-14] However, ZFN constructs are difficult to design and to construct for application in various organisms including plants. ZFNs also can suffer from imprecise DNA sequence recognition and are high in cost.[15,16] Later, a more promising tool, the TALENs derived from *Xanthomonas* bacteria, was developed and used for gene editing in plants.[17-21] But the complicated tandem repeat feature in the DNA binding domains of the TALEN protein makes the construction of TALEN genes tedious and complicated. The CRISPR/Cas9 system is the newest genome editing tool that emerged soon after the TALEN system was developed.

1.2 Comparison Among ZFNs, TALENs, and CRISPR/Cas9 Genome Editing Systems

ZFNs and TALENs are both chimeric nucleases comprised of a *Fok*I cleavage domain and a DNA-binding domain and function as dimers. With the complicated assembly of DNA-binding domain expression cassettes, poorer precision, and lower cutting efficiency of targets, ZFN and TALEN gene editing systems are easily surpassed by the CRISPR/Cas9 system.[22] In essence, the CRISPR/Cas9 system is quite simple. It has only two major components: the Cas9 protein and a single guide RNA (sgRNA).[23,24] The 20 nucleotides at the 5′ end of a sgRNA, as a component of the Cas9/sgRNA complex, can precisely pair by hybridization with DNA in a specific genome target site. The target site must be located immediately upstream of a protospacer adjacent motif (PAM; NGG for SpCas9 from *Streptococcus pyogenes*). This precise binding causes Cas9 to activate its two nuclease domains and produce a DSB in the target sequence.[24,25] Importantly, multiple sgRNAs with different target sequences can be designed for simultaneous multiplex editing.[26] Therefore, CRISPR/Cas9 is much easier to manipulate and has higher efficiency in inducing targeted mutations than ZFNs and TALENs. Indeed, the CRISPR/Cas9 system has become an essential genome editing technology and represents a true breakthrough in genome manipulation.[22,27] Recently, another CRISPR/Cas9-like system, CISPR/Cpf1, has been developed[28] and also successfully applied to genome editing in plants.[29-32]

Here, we introduce the CRISPR/Cas9-based gene editing system and its applications in plants.

2. THE MECHANISM OF THE CRISPR/CAS9 NUCLEASE SYSTEM

The *spCas9* gene is derived from the type II prokaryotic CRISPR adaptive immune system and contains a full-length coding region of 4107 bp.[24] The Cas9 protein contains separate RucV and HNH nuclease domains. Each nuclease domain has cleavage activity against one of the target DNA strands and, together, simultaneously produces a blunt-ended DSB.[24,33] The sgRNA is an artificial fusion of two separate RNAs essential for CRISPR activity, the protospacer-matching CRISPR RNA (crRNA), and the transactivating crRNA (tracrRNA).[23,24,34] The sgRNA structure closely mimics the original crRNA–tracrRNA duplex structure but, because it is encoded by a single gene, greatly simplifies manipulation of the CRISPR/Cas9 system. The target recognition sequence (not including PAM) is located at the 5′ end of the sgRNA and is responsible for specific hybridization to the chromosomal DNA target site.[23,24,35]

The Cas9 nuclease and the sgRNA form a Cas9 ribonucleo protein complex.[24,36] Vitally, the precise specificity of the Cas9/sgRNA nuclease complex for its target site is achieved by the hybridization of the sgRNA to the target, unlike ZFNs and TALENs that depend on the unique protein structure of their DNA binding domains for target recognition.[24] Both in vitro and in vivo, the Cas9/sgRNA complex can find its genomic target DNA sequence only when the target sequence is immediately adjacent to an NGG PAM sequence, indicating that PAM is essential for target recognition.[37] Once the target site is located, the Cas9/sgRNA complex separates the target DNA strands to facilitate sgRNA pairing with the target complementary strand. The RucV and HNH domains of Cas9 complex then, respectively, cut one of the two DNA strands three bases upstream of PAM, creating blunt-ended DSB. Then, the DSBs can be repaired either by NHEJ or by HDR to produce mutations in the targeted site. Multiple sgRNAs containing different targeting sequences can bind to different Cas9 proteins and simultaneously guide the various Cas9/sgRNA complexes to the corresponding target genes in the same cell.[23] Simultaneous editing greatly facilitates functional studies of multiple members of gene families and mutation of functionally related genes that control complex traits.[26,38]

The application of CRISPR/Cas9 system for gene editing was established first in human cells[23,24,35,39] and then applied to plants. Shortly thereafter, many CRISPR/Cas9 vector systems with improved editing efficiencies were developed for plants.[26,38,40–42] So far, CRISPR/Cas9 vector systems have been used to produce gene knockouts, DNA deletions, disruption of *cis*-regulatory elements, gene replacements (knock-ins), and suppression of virus infection.[43]

2.1 Structure of sgRNA/Cas9 Expression Cassettes

Starting with a 5′ 20-nt target recognition sequence, a standard sgRNA contains 98 nucleotides in total and plays a guiding role for the Cas9/sgRNA nuclease complex.[44] U3 or U6 small nuclear RNA gene promoters, which recruit RNA polymerase III, are generally used to drive transcription of sgRNAs in vivo.[45–48] In eukaryotes, the Cas9 protein requires attachment of nuclear localization signals for transport of Cas9 into the nucleus. To improve the expression of *Cas9* in plants, plant codon biases are incorporated into most *Cas9* genes modified for use in plant genome editing.[26,40,46,47,49–54] Several investigators have shown that the *Cas9* genes that use codon optimization for nonplant species have relatively lower editing efficiencies in plants than those using the plant-optimized codons.[45,46,55–57] For example, the codon-optimized *Cas9* gene (*Cas9p*) with a higher GC content in the 5′ terminal region, which mimics Gramineae genes,[58] produces Cas9 protein with higher editing efficiency in rice than noncodon-optimized Cas9.[26] In general, various constitutive promoters, such as those from the *Ubiquitin* genes of maize, rice, and *Arabidopsis* [and the 35S promoter from cauliflower mosaic virus (CaMV)] are used to drive the *Cas9* gene in monocot and dicot plants. In many cases, the *Ubiquitin* gene promoters show higher editing efficiencies than the CaMV 35S promoter in both monocot and dicot plants.[43,59]

2.2 Delivery of the *Cas9* and sgRNA Expression Cassettes into Plant Cells

For genome editing in plants, plasmids carrying the *Cas9* and sgRNA expression cassettes must be delivered into the plant cells. The *Cas9* cassettes and sgRNA cassettes can be arranged in separate constructs or single constructs.[45–47,50,55] In addition, the Cas9 protein can also be delivered in the form of RNA or ribonucleo proteins using biolistic delivery.[53] Woo *et al.* reported a DNA-free genome editing system for plants, which can generate transgene-free genome-edited lines by transfecting preassembled complexes of purified Cas9 protein and synthesized sgRNA into plant protoplasts.

This approach is particularly useful for vegetatively propagated plants.[60] It is also feasible to deliver synthesized sgRNA by biolistic transformation directly into plant cells previously transformed with a *Cas9* gene to induce targeted mutations.[53] Shan *et al.* used biolistic transformation to integrate the *Cas9* and sgRNA expression constructs into calli or immature embryos and obtained transgenic plants carrying heritable mutations.[50]

So far, most applications of CRISPR/Cas9 in plants use the *Agrobacterium*-mediated stable transformation system which integrates T-DNAs carrying both the *Cas9* and sgRNA expression cassettes into the plant genomes. In monocot and dicot plants, such as rice, maize, tobacco, tomato, potato, and poplar, the receptor tissues of *Agrobacterium*-mediated stable transformation are generally calli, immature embryos, or other isolated tissues. In contrast, *Agrobacterium*-mediated floral dip transformation is performed to produce genome editing in *Arabidopsis* and closely related brassica oil seed crops such as Camelina.[61] In addition, a sgRNA expression cassette carried by the tobacco rattle virus DNA can be introduced into *Cas9*-transformed tobacco using an agroinfiltration approach.[53,62,63]

2.3 CRISPR/Cas9 Vector Systems for Multiplex Genome Editing in Plants

The sgRNA expression cassettes are generally small in size (300–600 bp). Thus, single or multiple sgRNA cassettes can be prepared through target-adaptor ligation or overlapping PCR.[26,47,48,55,56,59] Then the prepared sgRNA expression cassttes can be cloned into a CRISPR/Cas9 binary vector directly. Another system utilizes self-cleavage ribozymes to produce functional and specific sgRNAs. These gene constructs can be transcribed by RNA polymerase II using constitutive, tissue-specific, or inducible promoters.[64]

Simultaneous editing of multiple genomic sites is useful in plant research, such as studies of multiple related genes, knockout of functionally redundant genes, or for guiding simultaneous genetic improvement of multiple traits in crop breeding. As mentioned earlier, *Agrobacterium*-mediated transformation is the main method to obtain stable transgenic plants in CRISPR/Cas9-based genome editing. However, it is a tricky and uncontrollable process to cotransform multiple T-DNAs from separate binary vectors into the same plant cells using *Agrobacterium*.[65] Therefore, several strategies have been developed to assemble multiple sgRNA expression cassettes into single CRISPR/Cas9 binary constructs. Several sgRNA expression cassettes containing different target sequences can be cloned into a binary vector sequentially using multiple rounds of cloning.[47,52,56,66] Alternatively, some researchers utilize multiple restriction enzymes that can yield sequential compatible palindromic sticky ends to allow insertion of a few (usually up to three) sgRNA cassettes simultaneously.[42,67] These traditional cloning methods are time-consuming, and researchers can clone only a few sgRNA expression cassettes into each CRISPR/Cas9 binary vector.

The Golden Gate cloning method can simultaneously and efficiently link multiple DNA fragments into one binary vector with a defined arrangement. In this method, distinctive type IIS restriction enzymes, such as *Bsa*I, are used to generate sequential compatible, nonpalindromic sticky ends among multiple DNA fragments.[68] Based on the Golden Gate cloning strategy, two CRISPR/Cas9 vector systems are developed to construct a binary construct with multiple sgRNA expression cassettes in a single-round cloning process.[26,40] Another CRISPR/Cas9 system uses an intermediate vector to clone multiple sgRNA cassettes. Then the fragments are recovered and ligated into a binary vector by Golden Gate ligation.[69] Alternatively, with the concerted use of T5 exonuclease, Phusion DNA polymerase, and *Taq* DNA ligase, the Gibson assembly method can be used to efficiently join multiple DNA fragments containing homologous termini.[70] This method also is used to ligate multiple sgRNA expression cassettes into a CRISPR/Cas9 binary vector in a single reaction.[26]

A polycistronic tRNA–gRNA system has been developed to generate multiple sgRNAs, by using tRNA precursor sequences (pre-RNA) to link the sgRNA sequences.[38] The polycistronic pre-RNA/sgRNA sequences can be transcribed using a U3 or U6 promoter and processed in vivo to releases multiple sgRNAs for multiplex genome targeting in plants.[38] However, the upper limit in terms of number of sgRNA units that can be accommodated within a polycistronic pre-tRNA/sgRNA and that can be successfully processed after transcription by a U3 or U6 promoter remains unclear.

Our CRISPR/Cas9 vector system for multiplex genome editing in plants consists of 5 binary vectors and 12 sgRNA vectors.[26,59] The *Cas9p* gene in the binary vectors is driven by a maize *Ubiquitin* promoter or a 35S promoter. This system uses target-adaptor ligation/PCR or overlapping PCR, with the sgRNA vectors as the templates, to amplify fragments of sgRNA expression cassettes. Then the PCR-generated cassette fragments are directly and simultaneously ligated into a selected binary vector by a single round of Golden Gate or Gibson assembly. Using the Golden Gate method, we prepared a CRISPR/Cas9 binary construct containing eight sgRNA expression cassettes and used it to mutate eight genes simultaneously in rice.[26]

2.4 Efficiency and Features of CRISPR/Cas9-Induced Mutations in Rice

At present, genome editing has been successfully performed in several plant species using the CRISPR/Cas9 system, and stable genetic mutants have been obtained.[26,40,50,53,54,56,71-73] However, the editing efficiencies of different CRISPR/Cas9 systems for the same species can be very different. Moreover, the editing efficiency varies when the system is applied in different plant species.[43]

Many factors may affect the efficiency and outcomes of CRISPR/Cas9 genome editing in plants, such as the vector system (the *Cas9* gene and the promoters used to drive the *Cas9* and sgRNA genes), target sites, targeted plants, and transformation strategies. Here, we focus on the efficiency and outcomes of CRISPR/Cas9 genome editing in rice.

2.5 Outcomes Using the CRISPR/Cas9 Vector System in Rice

Based on our research, about 86% on average of targeted mutations can be obtained in T_0 transgenic rice plants; about half of these mutants are biallelic (two distinct allelic mutations), and the remaining mutations are homozygous or heterozygous, with rare chimeric mutations.[26,74] CRISPR/Cas9-induced mutations are heritable in subsequent generations.[26,52,74,75] In general, uniform mutations can be transmitted into the progeny more reliably than chimeric mutations.[43]

It is noteworthy that a higher proportion of homozygous mutations occurs in T_0 rice plants than expected, which reached as high as 27.8% of total mutations.[26,43] It is proposed that two mechanisms may be involved in producing homozygous mutations. First, two independent identical mutations at both homologous chromosomal sites may be induced by NHEJ, especially for the high-frequency "A" and "T" base insertions. In the second senario, NHEJ creates the first mutated allele, and then the mutated allele serves as a template to produce the same mutation in the opposite allele by HDR.[26] This second type of NHEJ–HDR repair, which increases the proportion of homozygous mutations, has been reported previously in genome editing research with animals.[76-78]

Several transient expression assays showed that sgRNA concentration can be the limiting factor for editing efficiency.[39,47] However, our previous research indicated that different U3 and U6 promoters resulted invaried sgRNA expression levels in rice, but produced similar editing efficiencies, suggesting that in most cases the sgRNA level is sufficient to function well with the existing concentration of Cas9 protein.

2.6 Characterization of Targeted Editing in Transgenic Rice

As mentioned earlier, the Cas9/sgRNA nuclease complex creates blunt-ended DSB three bases upstream of the PAM site. Based on reported cases, about half of CRISPR/Cas9-induced mutations are single-base insertions (mostly A and T), and the rest are small deletions, ranging from 1 to 50 bp, while base substitution and two or more base insertions are rare.[26,74,75] Therefore, targeting the coding regions of genes has a high possibility of causing a frame-shift and, as a consequence, loss-of-function mutations. If two or more sites are selected within a chromosomal region, a fragment deletion (up to hundreds of kilobase pairs in length) can occur between the target sites.[26,47,52,55,56,79] This feature can be applied to delete an entire gene or certain chromosomal regions containing multiple genes.

2.7 Factors Affecting the Editing Efficiency of CRISPR/Cas9-Based Gene Editing in Plants

The efficiencies of CRISPR/Cas9-mediated genome editing vary dramatically in plants.[80] It has been shown that the codon optimization of *Cas9*, the promoter used to drive the expression of *Cas9*, and the position effect of the T-DNA insertions may directly affect *Cas9* expression levels and the targeting efficiencies.[26,56,66,67,81]

The GC content of the target sequence or other sequence compositions may also affect the CRISPR/Cas9-induced editing efficiency.[26,74] In rice, high GC contents (50%–70%) of the target sequences produce relatively higher editing efficiencies.[26] The specific transcription initiation site for the U3 promoters for sgRNA expression is an "A" nucleotide. Whereas it is a G nucleotide for U6 promoters. In early studies, the first nucleotide of the targets was restricted to A or G, by which the regular forms are AN19NGG matching the U3 promoters and GN19NGG matching the U6 promoters.[35,39] However, it has been shown that one or a few extra nucleotides (irregular sgRNAs) at the 5′ end of sgRNAs do not influence the genome editing efficiency of Cas9/sgRNA nuclease complex.[26,55,82] Hence, target selections are not necessarily limited by the initial nucleotides, thus broadening the possibilities for target selection.[26] Furthermore, if the target sequence forms a stem–loop structure with the sgRNA scaffold sequence (i.e., the portion of the sgRNA binding to Cas9), the editing efficiency may be decreased.[26] Therefore, genomic sites with more than six continuous nucleotides paired with the sgRNA scaffold sequence should be avoided.[26]

In addition, the repairing competence among various plants or diverse tissues of same plant may affect the genomic editing efficiency of the Cas9/sgRNA nuclease complex. The DNA DSBs caused by the CRISPR/Cas9 system can be repaired correctly in a species with high DNA repair competence and may result in few, if any, mutations.[83]

3. ANALYSIS OF TARGETED MUTATIONS INDUCED BY CRISPR/CAS9 IN PLANTS

To validate a newly established CRISPR/Cas9 vector system or before applying an established CRISPR/Cas9 system to a new plant species, one of the critical considerations is the editing efficiency. In addition, it is necessary to confirm the sequences of CRISPR/Cas9-induced mutations before further study. Several approaches have been developed to verify the nature of targeted mutations.[46,51,75,84–86]

3.1 Approaches to Verify Targeted Mutations

If properly designed, a restriction enzyme site-based assay can be used to detect Cas9/sgRNA-mediated mutation events. If the target sequence contains a restriction enzyme site, the Cas9 cleavage and error-prone NHEJ DNA repair may produce targeted mutations and destroy the restriction enzyme site. Is so, it is simple to enrich mutated sequences by restriction enzyme digestion of the genomic DNA templates or PCR amplicons.[8,84] This strategy has been exploited to determine the presence of targeted mutations and measure the editing efficiency in several cases.[45,46,50,55] However, this method requires a presence of a restriction enzyme site within the target core sequences. The Surveyor nuclease and T7 Endonuclease I assays also can be applied in a similar way. This method utilizes the ability of T7 endonuclease I to cleave PCR-produced hybrid DNA fragments that contain unpaired nucleotides.[87] This method can be used to confirm the targeted mutations and calculate the mutagenesis efficiency. However, this method has lower detection sensitivity than the restriction enzyme site-based assay but, importantly, is suitable for any target sequence.[84]

The single-strand conformation polymorphism method is also used to determine targeted mutations induced by CRISPR/Cas9.[85] The principle of this method is that single-strand DNAs with nucleotide variations can change their conformations and exhibit differential migration rates in a non-denaturing PAGE gel. Another PAGE-based method can detect heteroduplex DNAs with targeted mutations.[88]

When mutations occur, mutated sequences may posses a different melting temperature than the wild-type sequence. Therefore, targeted mutations in PCR amplicons can be detected by a high-resolution melting assay,[51,89] but with low detection sensitivity. However, all these methods described earlier can only detect the presence of mutations but cannot determine mutated nucleotide sequences.

In addition, high-throughput sequencing (deep sequencing) of the whole genome or single/multiple PCR amplicons is suitable for detecting rare (low frequency) mutations and complicated chimeric mutations, especially for identifying possible off-target mutations over the whole genome.[51,75] However, this method is costly and time-consuming.

3.2 A Method, DSD, for Decoding Superimposed Sequencing Chromatograms

Several researches have shown that CRISPR/Cas9-based genome editing is highly efficient and can produce uniform mutations in high proportions (including biallelic, homozygous, or heterozygous mutations) in the first transgenic generation (T_0) for rice or other plant species.[26,52,74] In order to determine the sequences of the targeted mutations, cloning of the PCR amplicons containing targeted site(s) and sequencing of multiple clones (at least six clones for uniform mutations) are required. However, this conventional method is costly, tedious, and time-consuming. If such PCR amplicons are directly sequenced, heterozygous and biallelic mutations will produce overlapping peaks (double traces) in the sequencing chromatograms beginning from the mutation site. To resolve this problem, we developed a method called Degenerate Sequence Decoding (DSD) for decoding of such superimposed sequencing chromatograms from direct sequencing of PCR products with heterozygous and biallelic mutations.[86] Manual decoding of a sequencing file using the DSD method needs only about 3–4 min. This method was used to decode hundreds of targeted mutation events in rice and *Arabidopsis*.[26]

3.3 A Software DSDecode Based on the DSD Method

The manual operation of the DSD method is time-consuming when dealing with large numbers of superimposed sequencing chromatograms. Therefore, based on the principle of the DSD method, we developed a web-based software tool, DSDecode (http://dsdecode.scgene.com/).[90] This tool can rapidly and automatically decode superimposed and regular sequencing chromatograms (ab1 format) of PCR amplicons containing various types of targeted mutations and is user-friendly. Recently, we updated and released a new version of this web tool, DSDecodeM (http://skl.scau.edu.cn/), which enables a person to decode multiple sequencing files simultaneously and more rapidly. A PC version of this tool, DSDecodeML, is also available upon requested. Using this tool, the cost and time for analysis of genome editing in diploid organisms can be greatly decreased, promoting the applications of targeted genome editing in various organism.

4. APPLICATIONS OF CRISPR/CAS9 SYSTEM FOR PLANT FUNCTIONAL STUDIES AND CROP IMPROVEMENT

The robust CRISPR/Cas9-based genome editing system is becoming a powerful tool for genetic improvement of important traits in crops, such as yield, plant architecture, nutrient content, disease resistance, and adaption to stresses.[43] In some cases, a simple knockout of a negative regulatory gene can improve a particular trait. For example, knockout of all of the three *TaMLO* homologs simultaneously in hexaploid wheat produced resistance to powdery mildew.[71] Gene editing of certain quantitative trait loci improved rice grain weight.[109] Precise genomic DNA modification at the *ARGOS8* locus was shown to improve maize grain yield under field drought stress conditions.[91]

It has been shown that phenotypic mutants could be obtained in the T_0 generation of woody plants using CRISPR/Cas9 editing. This is especially useful in breeding of woody plants because of their very long reproductive cycles.[92,93] Moreover, the CRISPR/Cas9 system has been adapted for use as an antivirus tool that has been utilized to specifically cleave a particular viral DNA so that DNA virus infection was repressed.[94–97]

4.1 Use of the CRISPR/Cas9 Vector System From Our Lab

Due to its high efficiency and easy use, our CRISPR/Cas9 system has been used in many laboratories for basic and applied research with various plants.

We are using this system to study the functions of large number of new genes in rice, as well as to improve certain agronomic traits of rice. Likewise, other laboratories have reported their results in rice molecular breeding or functional study using our CRISPR/Cas9 system. For example, targeted mutation of the *TMS5* gene in rice cultivars led to rapid development of temperature-sensitive lines for use in hybrid rice production.[7] Wang *et al.* edited the ERF transcription factor gene, *OsERF922*, which is a negative regulator, to enhance rice disease resistance.[98] Ma *et al.* knocked out the *OsGAN1* gene in rice and proved that this gene positively regulated plant height and root length.[99] Using gene knockout, the *OsABCG26* gene was shown to be required in formation of anther cuticle and pollen exine in rice,[100] and *OsTCD10* was shown to play an important role in chloroplast development under cold stress.[101] Fan *et al.* expanded use of our system to demonstrate that CRISPR/Cas9 can be effectively employed to create knockout mutations in woody plants in the T_0 generation.[92]

5. CONCLUDING REMARKS AND PERSPECTIVES

Development of the CRISPR/Cas9 genome editing technology will have a revolutionary influence on gene function research in plants and for crop genetic improvement.[83,102,103] Compared to medical and clinical research, plant genome editing does not involve in ethical issues. Thus, the CRISPR/Cas9 genome editing technology is more immediately suitable for applied research in plants. Technically, CRISPR/Cas9-induced mutations are substantially equal to those generated by random mutagenesis, but with much higher specificity and efficiency. In addition, with segregation of genes between progeny during breeding, the transgenic T-DNA region harboring the *Cas9* and sgRNA genes and the antibiotic-resistant marker genes can be simply excluded, creating transgene-free gene-edited plant lines.[52,57]

Although the CRISPR/Cas9 genome editing system has shown great potential in gene function and applied research, this technology remains far from fully explored. Although several CRISPR/Cas9 vector platforms have been established for genome editing in plants, the efficiency, specificity, and range of accessible targets need to be improved. However, precise gene editing in plants, including DNA fragment knock-in and gene replacement, remains a major challenge.[43] Furthermore, Cas9 variants modified to contain gene activation or repression domains can be used to regulate the expression of targeted genes.[104] Indeed, a CRISPR/Cas9-based toolbox for gene activation and repression in plants has been developed.[69] The CRISPR/Cas9 and similar systems may be adapted for many more novel applications, such as RNA cleavage,[105] epigenomic regulation,[106,107] and chromatin imaging.[108] With the development of CRISPR/Cas9 and other emerging genome editing systems, a revolutionary change is currently happening for biology research and crop molecular breeding.

REFERENCES

1. Kuromori T, Wada T, Kamiya A, et al. A trial of phenome analysis using 4000 Ds-insertional mutants in gene-coding regions of Arabidopsis. *Plant J.* 2006;47:640–651.
2. Wu CY, Li XJ, Yuan WY, et al. Development of enhancer trap lines for functional analysis of the rice genome. *Plant J.* 2003;35:418–427.
3. Yang Y, Li Y, Wu C. Genomic resources for functional analyses of the rice genome. *Curr Opin Plant Biol.* 2013;16:157–163.
4. McCallum CM, Comai L, Greene EA, et al. Targeted screening for induced mutations. *Nat Biotechnol.* 2000;18:455–457.
5. Symington LS, Gautier J. Double-strand break end resection and repair pathway choice. In: Bassler BL, Lichten M, Schupbach G, eds; 2011:247–271. Annual Review of Genetics; vol 45.
6. Xu R, Li H, Qin R, et al. Generation of inheritable and "transgene clean" targeted genome-modified rice in later generations using the CRISPR/Cas9 system. *Sci Rep.* 2015;5:11491.
7. Zhou H, He M, Li J, et al. Development of commercial thermo-sensitive genic male sterile rice accelerates hybrid rice breeding using the CRISPR/Cas9-mediated TMS5 editing system. *Sci Rep.* 2016;6:37395.
8. Lloyd A, Plaisier CL, Carroll D, et al. Targeted mutagenesis using zinc-finger nucleases in Arabidopsis. *Proc Natl Acad Sci USA.* 2005;102:2232–2237.
9. Zhang F, Maeder ML, Unger-Wallace E, et al. High frequency targeted mutagenesis in Arabidopsis thaliana using zinc finger nucleases. *Proc Natl Acad Sci USA.* 2010;107:12028–12033.
10. Zhang F, Voytas DF. Targeted mutagenesis in Arabidopsis using zinc-finger nucleases. In: Birchler JA, ed. *Methods in Molecular Biology*; 2011:167–177. vol 701.
11. Petolino JF. Genome editing in plants via designed zinc finger nucleases. *In Vitro Cell Dev Biol Plant.* 2015;51:1–8.
12. Kumar V, Jain M. The CRISPR-Cas system for plant genome editing: advances and opportunities. *J Exp Bot.* 2015;66:47–57.
13. Smith J, Bibikova M, Whitby FG, et al. Requirements for double-strand cleavage by chimeric restriction enzymes with zinc finger DNA-recognition domains. *Nucleic Acids Res.* 2000;28:3361–3369.
14. Bibikova M, Golic M, Golic KG, et al. Targeted chromosomal cleavage and mutagenesis in Drosophila using zinc-finger nucleases. *Genetics.* 2002;161:1169–1175.
15. Ramirez CL, Foley JE, Wright DA, et al. Unexpected failure rates for modular assembly of engineered zinc fingers. *Nat Methods.* 2008;5:575 [vol 5, p. 374, 2008].
16. Sanjana NE, Cong L, Zhou Y, et al. A transcription activator-like effector toolbox for genome engineering. *Nat Protoc.* 2012;7:171–192.
17. Christian M, Cermak T, Doyle EL, et al. Targeting DNA double-strand breaks with TAL effector nucleases. *Genetics.* 2010;186:476–757.

18. Boch J, Bonas U. Xanthomonas AvrBs3 family-type III effectors: discovery and function. In: NK VA, Bruening G, Leach JE, eds; 2010:419–436. Annual Review of Phytopathology; vol 48.
19. Boch J, Scholze H, Schornack S, et al. Breaking the code of DNA binding specificity of TAL-type III effectors. *Science*. 2009;326:1509–1512.
20. Moscou MJ, Bogdanove AJ. A simple cipher governs DNA recognition by TAL effectors. *Science*. 2009;326:1501.
21. Li T, Huang S, Jiang WZ, et al. TAL nucleases (TALNs): hybrid proteins composed of TAL effectors and FokI DNA-cleavage domain. *Nucleic Acids Res*. 2011;39:359–372.
22. Ledford H. CRISPR, the disruptor. *Nature*. 2015;522:20–24.
23. Cong L, Ran FA, Cox D, et al. Multiplex genome engineering using CRISPR/Cas systems. *Science*. 2013;339:819–823.
24. Jinek M, Chylinski K, Fonfara I, et al. A programmable dual-RNA-guided DNA endonuclease in adaptive bacterial immunity. *Science*. 2012;337:816–821.
25. Doudna JA, Charpentier E. The new frontier of genome engineering with CRISPR-Cas9. *Science*. 2014;346:1077.
26. Ma X, Zhang Q, Zhu Q, et al. A robust CRISPR/Cas9 system for convenient, high-efficiency multiplex genome editing in monocot and dicot plants. *Mol Plant*. 2015;8:1274–1284.
27. Ledford H. Riding the CRISPR wave. *Nature*. 2016;531:156–159.
28. Zetsche B, Gootenberg JS, Abudayyeh OO, et al. Cpf1 is a single RNA-guided endonuclease of a class 2 CRISPR-Cas system. *Cell*. 2015;163:759–771.
29. Xu R, Qin R, Li H, et al. Generation of targeted mutant rice using a CRISPR-Cpf1 system. *Plant Biotechnol J*. 2016. http://dx.doi.org/10.1111/pbi.12669.
30. Endo A, Masafumi M, Kaya H, et al. Efficient targeted mutagenesis of rice and tobacco genomes using Cpf1 from Francisella novicida. *Sci Rep*. 2016;6:38169.
31. Hu X, Wang C, Liu Q, et al. Targeted mutagenesis in rice using CRISPR-Cpf1 system. *J Genet Genomics*. 2017;44:71–73.
32. Tang X, Lowder LG, Zhang T, et al. A CRISPR-Cpf1 system for efficient genome editing and transcriptional repression in plants. *Nat Plants*. 2017;3:17018.
33. Villion M, Moineau S. The double-edged sword of CRISPR-Cas systems. *Cell Res*. 2013;23:15–17.
34. Mali P, Esvelt KM, Church GM. Cas9 as a versatile tool for engineering biology. *Nat Methods*. 2013;10:957–963.
35. Mali P, Yang L, Esvelt KM, et al. RNA-guided human genome engineering via Cas9. *Science*. 2013;339:823–826.
36. Jiang F, Zhou K, Ma L, et al. A Cas9-guide RNA complex preorganized for target DNA recognition. *Science*. 2015;348:1477–1481.
37. Sternberg SH, Redding S, Jinek M, et al. DNA interrogation by the CRISPR RNA-guided endonuclease Cas9. *Nature*. 2014;507:62.
38. Xie K, Minkenberg B, Yang Y. Boosting CRISPR/Cas9 multiplex editing capability with the endogenous tRNA-processing system. *Proc Natl Acad Sci USA*. 2015;112:3570–3575.
39. Jinek M, East A, Cheng A, et al. RNA-programmed genome editing in human cells. *Elife*. 2013;2:e00471.
40. Xing H, Dong L, Wang Z, et al. A CRISPR/Cas9 toolkit for multiplex genome editing in plants. *BMC Plant Biol*. 2014;14:327.
41. Wang C, Shen L, Fu Y, et al. A simple CRISPR/Cas9 system for multiplex genome editing in rice. *J Genet Genomics*. 2015;42:703–706.
42. Zhang Z, Mao Y, Ha S, et al. A multiplex CRISPR/Cas9 platform for fast and efficient editing of multiple genes in Arabidopsis. *Plant Cell Rep*. 2016;35:1519–1533.

43. Ma X, Zhu Q, Chen Y, et al. CRISPR/Cas9 platforms for genome editing in plants: developments and applications. *Mol Plant.* 2016;9:961–974.
44. Nishimasu H, Ran FA, Hsu PD, et al. Crystal structure of cas 9 in complex with guide RNA and target DNA. *Cell.* 2014;156:935–949.
45. Nekrasov V, Staskawicz B, Weigel D, et al. Targeted mutagenesis in the model plant Nicotiana benthamiana using Cas9 RNA-guided endonuclease. *Nat Biotechnol.* 2013;31:691–693.
46. Jiang W, Zhou H, Bi H, et al. Demonstration of CRISPR/Cas9/sgRNA-mediated targeted gene modification in Arabidopsis, tobacco, sorghum and rice. *Nucleic Acids Res.* 2013;41:e188.
47. Li J, Norville JE, Aach J, et al. Multiplex and homologous recombination-mediated genome editing in Arabidopsis and Nicotiana benthamiana using guide RNA and Cas9. *Nat Biotechnol.* 2013;31:688–691.
48. Shan Q, Gao C. Research progress of genome editing and derivative technologies in plants. *Yi Chuan.* 2015;37:953–973.
49. Miao J, Guo D, Zhang J, et al. Targeted mutagenesis in rice using CRISPR-Cas system. *Cell Res.* 2013;23:1233–1236.
50. Shan Q, Wang Y, Li J, et al. Targeted genome modification of crop plants using a CRISPR-Cas system. *Nat Biotechnol.* 2013;31:686–688.
51. Fauser F, Schiml S, Puchta H. Both CRISPR/Cas-based nucleases and nickases can be used efficiently for genome engineering in Arabidopsis thaliana. *Plant J.* 2014;79:348–359.
52. Zhou H, Liu B, Weeks DP, et al. Large chromosomal deletions and heritable small genetic changes induced by CRISPR/Cas9 in rice. *Nucleic Acids Res.* 2014;42:10903–10914.
53. Svitashev S, Young JK, Schwartz C, et al. Targeted mutagenesis, precise gene editing, and site-specific gene insertion in maize using cas9 and guide RNA. *Plant Physiol.* 2015;169:931–945.
54. Gao J, Wang G, Ma S, et al. CRISPR/Cas9-mediated targeted mutagenesis in Nicotiana tabacum. *Plant Mol Biol.* 2015;87:99–110.
55. Xie K, Yang Y. RNA-guided genome editing in plants using a CRISPR-Cas system. *Mol Plant.* 2013;6:1975–1983.
56. Mao Y, Zhang H, Xu N, et al. Application of the CRISPR-Cas system for efficient genome engineering in plants. *Mol Plant.* 2013;6:2008–2011.
57. Lawrenson T, Shorinola O, Stacey N, et al. Induction of targeted, heritable mutations in barley and Brassica oleracea using RNA-guided Cas9 nuclease. *Genome Biol.* 2015;16:258.
58. Wong G, Wang J, Tao L, et al. Compositional gradients in Gramineae genes. *Genome Res.* 2002;12:851–856.
59. Ma X, Liu YG. CRISPR/Cas9-based multiplex genome editing in monocot and dicot plants. *Curr Protoc Mol Biol.* 2016;115:31–36.
60. Woo JW, Kim J, Il Kwon S, et al. DNA-free genome editing in plants with preassembled CRISPR-Cas9 ribonucleoproteins. *Nat Biotechnol.* 2015;33:1156–1162.
61. Jiang WZ, Henry IM, Lynagh PG, et al. Significant enhancement of fatty acid composition in seeds of the allohexaploid, Camelina sativa, using CRISPR/Cas9 gene editing. *Plant Biotechnol J.* 2017;15:648–657.
62. Ali Z, Abul-faraj A, Li L, et al. Efficient virus-mediated genome editing in plants using the CRISPR/Cas9 system. *Mol Plant.* 2015;8:1288–1291.
63. Yin K, Han T, Liu G, et al. A geminivirus-based guide RNA delivery system for CRISPR/Cas9 mediated plant genome editing. *Sci Rep.* 2015;5:14926.
64. Gao Y, Zhao Y. Self-processing of ribozyme-flanked RNAs into guide RNAs in vitro and in vivo for CRISPR-mediated genome editing. *J Integr Plant Biol.* 2014;56:343–349.

65. Hiei Y, Komari T. Agrobacterium-mediated transformation of rice using immature embryos or calli induced from mature seed. *Nat Protoc*. 2008;3:824–834.
66. Yan L, Wei S, Wu Y, et al. High-efficiency genome editing in Arabidopsis using YAO promoter-driven CRISPR/Cas9 system. *Mol Plant*. 2015;8:1820–1823.
67. Wang Z, Xing H, Dong L, et al. Egg cell-specific promoter-controlled CRISPR/Cas9 efficiently generates homozygous mutants for multiple target genes in Arabidopsis in a single generation. *Genome Biol*. 2015;16:144.
68. Engler C, Kandzia R, Marillonnet S. A one pot, one step, precision cloning method with high throughput capability. *PLoS One*. 2008;3:e3647.
69. Lowder LG, Zhang D, Baltes NJ, et al. A CRISPR/Cas9 toolbox for multiplexed plant genome editing and transcriptional regulation. *Plant Physiol*. 2015;169:971.
70. Gibson DG, Young L, Chuang R, et al. Enzymatic assembly of DNA molecules up to several hundred kilobases. *Nat Methods*. 2009;6:341–343.
71. Wang Y, Cheng X, Shan Q, et al. Simultaneous editing of three homoeoalleles in hexaploid bread wheat confers heritable resistance to powdery mildew. *Nat Biotechnol*. 2014;32:947–951.
72. Brooks C, Nekrasov V, Lippman ZB, et al. Efficient gene editing in tomato in the first generation using the clustered regularly interspaced short palindromic repeats/CRISPR-Associated9 system. *Plant Physiol*. 2014;166:1292–1297.
73. Wang S, Zhang S, Wang W, et al. Efficient targeted mutagenesis in potato by the CRISPR/Cas9 system. *Plant Cell Rep*. 2015;34:1473–1476.
74. Zhang H, Zhang J, Wei P, et al. The CRISPR/Cas9 system produces specific and homozygous targeted gene editing in rice in one generation. *Plant Biotechnol J*. 2014;12:797–807.
75. Feng Z, Mao Y, Xu N, et al. Multigeneration analysis reveals the inheritance, specificity, and patterns of CRISPR/Cas-induced gene modifications in Arabidopsis. *Proc Natl Acad Sci USA*. 2014;111:4632–4637.
76. Wu Y, Liang D, Wang Y, et al. Correction of a genetic disease in mouse via use of CRISPR-Cas9. *Cell Stem Cell*. 2013;13:659–662.
77. Slaymaker IM, Gao L, Zetsche B, et al. Rationally engineered Cas9 nucleases with improved specificity. *Science*. 2016;351:84–88.
78. Gantz VM, Bier E. The mutagenic chain reaction: a method for converting heterozygous to homozygous mutations. *Science*. 2015;348:442–444.
79. Zhao Y, Zhang C, Liu W, et al. An alternative strategy for targeted gene replacement in plants using a dual-sgRNA/Cas9 design. *Sci Rep*. 2016;6:23890.
80. Bortesi L, Fischer R. The CRISPR/Cas9 system for plant genome editing and beyond. *Biotechnol Adv*. 2015;33:41–52.
81. Mao Y, Zhang Z, Feng Z, et al. Development of germ-line-specific CRISPR-Cas9 systems to improve the production of heritable gene modifications in Arabidopsis. *Plant Biotechnol J*. 2016;14:519–532.
82. Ran FA, Hsu PD, Lin C, et al. Double nicking by RNA-guided CRISPR Cas9 for enhanced genome editing specificity. *Cell*. 2013;154:1380–1389 [2013; 155: 479–480].
83. Ma X, Liu Y. CRISPR/Cas9-based genome editing systems and the analysis of targeted genome mutations in plants. *Yi Chuan*. 2016;38:118–125.
84. Voytas DF. Plant genome engineering with sequence-specific nucleases. In: Merchant SS, ed; 2013:327–350. Annual Review of Plant Biology; vol 64.
85. Zheng X, Yang S, Zhang D, et al. Effective screen of CRISPR/Cas9-induced mutants in rice by single-strand conformation polymorphism. *Plant Cell Rep*. 2016;35: 1545–1554.
86. Ma X, Chen L, Zhu Q, et al. Rapid decoding of sequence-specific nuclease-induced heterozygous and biallelic mutations by direct sequencing of PCR products. *Mol Plant*. 2015;8:1285–1287.

87. Guschin DY, Waite AJ, Katibah GE, et al. A rapid and general assay for monitoring endogenous gene modification. In: Mackay JP, Segal DJ, eds. *Engineered Zinc Finger Proteins*. 2010:247–256. vol 649.
88. Zhu X, Xu Y, Yu S, et al. An efficient genotyping method for genome-modified animals and human cells generated with CRISPR/Cas9 system. *Sci Rep*. 2014;4.
89. Dahlem TJ, Hoshijima K, Jurynec MJ, et al. Simple methods for generating and detecting locus-specific mutations induced with TALENs in the zebrafish genome. *PLoS Genet*. 2012;8.
90. Liu W, Xie X, Ma X, et al. DSDecode: a web-based tool for decoding of sequencing chromatograms for genotyping of targeted mutations. *Mol Plant*. 2015;8:1431–1433.
91. Shi J, Gao H, Wang H, et al. ARGOS8 variants generated by CRISPR-Cas9 improve maize grain yield under field drought stress conditions. *Plant Biotechnol J*. 2017;15:207–216.
92. Fan D, Liu T, Li C, et al. Efficient CRISPR/Cas9-mediated targeted mutagenesis in populus in the first generation. *Sci Rep*. 2015;5:12217.
93. Tsai C, Xue L. CRISPRing into the woods. *GM Crops Food*. 2015;6:206–215.
94. Ji X, Zhang H, Zhang Y, et al. Establishing a CRISPR-Cas-like immune system conferring DNA virus resistance in plants. *Nat Plants*. 2015;1:15144.
95. Ali Z, Abulfaraj A, Idris A, et al. CRISPR/Cas9-mediated viral interference in plants. *Genome Biol*. 2015;16:238.
96. Baltes NJ, Hummel AW, Konecna E, et al. Conferring resistance to geminiviruses with the CRISPR-Cas prokaryotic immune system. *Nat Plants*. 2015;1:1–4.
97. Chaparro-Garcia A, Kamoun S, Nekrasov V. Boosting plant immunity with CRISPR/Cas. *Genome Biol*. 2015;16:254.
98. Wang F, Wang C, Liu P, et al. Enhanced rice blast resistance by CRISPR/Cas9-targeted mutagenesis of the ERF transcription factor gene OsERF922. *PLoS One*. 2016;11:e0154027.
99. Ma X, Feng F, Wei H, et al. Genome-wide association study for plant height and grain yield in rice under contrasting moisture regimes. *Front Plant Sci*. 2016;7:1801.
100. Chang Z, Chen Z, Yan W, et al. An ABC transporter, OsABCG26, is required for anther cuticle and pollen exine formation and pollen-pistil interactions in rice. *Plant Sci*. 2016;253:21–30.
101. Wu L, Wu J, Liu Y, et al. The rice pentatricopeptide repeat gene TCD10 is needed for chloroplast development under cold stress. *Rice*. 2016;9:67.
102. Belhaj K, Chaparro-Garcia A, Kamoun S, et al. Editing plant genomes with CRISPR/Cas9. *Curr Opin Biotechnol*. 2015;32:76–84.
103. Weeks DP, Spalding MH, Yang B. Use of designer nucleases for targeted gene and genome editing in plants. *Plant Biotechnol J*. 2016;14:483–495.
104. Gilbert LA, Larson MH, Morsut L, et al. CRISPR-mediated modular RNA-guided regulation of transcription in eukaryotes. *Cell*. 2013;154:442–451.
105. O'Connell MR, Oakes BL, Sternberg SH, et al. Programmable RNA recognition and cleavage by CRISPR/Cas9. *Nature*. 2014;516:263.
106. Kearns NA, Pham H, Tabak B, et al. Functional annotation of native enhancers with a Cas9-histone demethylase fusion. *Nat Methods*. 2015;12:401.
107. Hilton IB, D'Ippolito AM, Vockley CM, et al. Epigenome editing by a CRISPR-Cas9-based acetyltransferase activates genes from promoters and enhancers. *Nat Biotechnol*. 2015;33:225–510.
108. Chen B, Gilbert LA, Cimini BA, et al. Dynamic imaging of genomic loci in living human cells by an optimized CRISPR/Cas system. *Cell*. 2013;155:1479–1491.
109. Xu R, Yang Y, Qin R, et al. Rapid improvement of grain weight via highly efficient CRISPR/Cas9-mediated multiplex genome editing in rice. *J Genet Genomics*. 2016;43:529–532.

CHAPTER NINE

On Improving CRISPR for Editing Plant Genes: Ribozyme-Mediated Guide RNA Production and Fluorescence-Based Technology for Isolating Transgene-Free Mutants Generated by CRISPR

Yubing He*, Rongchen Wang*, Xinhua Dai†, Yunde Zhao†,[1]
*National Key Laboratory of Crop Genetic Improvement and National Center of Plant Gene Research (Wuhan), Huazhong Agricultural University, Wuhan, China
†Section of Cell and Developmental Biology, University of California San Diego, La Jolla, CA, United States
[1]Corresponding author: e-mail address: yundezhao@ucsd.edu

Contents

Abstract

CRISPR/Cas9-mediated genome editing technology has been used to successfully edit numerous genes in various organisms including plants. There are still two major challenges in using CRISPR/Cas9 technology for gene editing in plants. First, there are very limited choices of promoters that are suitable for in vivo production of single-guide RNAs (sgRNAs), which is complementary to the target sequence and which guides Cas9 to generate double-strand breaks at the target site. It is especially difficult to produce sgRNA molecules with temporal and spatial precision. Second, there is a lack of efficient methods for identifying plants that (1) contain heritable and stable mutations generated by CRISPR/Cas9, and (2) no longer harbor the CRISPR/Cas9 construct and other transgenes. In this chapter, we describe the development of a ribozyme-based

Progress in Molecular Biology and Translational Science, Volume 149
ISSN 1877-1173
http://dx.doi.org/10.1016/bs.pmbts.2017.03.012

strategy that enables the production of sgRNA molecules from any chosen promoter. More importantly, the ribozyme-based technology makes it feasible to produce sgRNAs with temporal and spatial precision, greatly expanding the scope and applications of CRISPR/Cas9 technology. We also developed a fluorescence-based technology that allows us to efficiently and reliably isolate *Cas9*-free stable Arabidopsis mutants. Thus, we provide effective protocols to overcome two important obstacles in using CRISPR/Cas9 for editing genes in plants.

The CRISPR/Cas9 system (clustered regularly interspaced short palindromic repeats/CRISPR-associated endonuclease Cas9) has become a dominant force in genome editing. The number of cases of successful CRISPR/Cas9-mediated gene editing has exploded exponentially since 2013 when the first in vivo gene editing by CRISPR/Cas9 was reported.[1–3] CRISPR/Cas9 technology has been quickly adopted for biological research and for crop improvement because of its simplicity and effectiveness. Cas9 is an endonuclease with two nuclease domains that can generate double-strand DNA breaks (DSBs) when it binds to a target DNA sequence.[3] However, Cas9 only binds to a target DNA sequence when Cas9 contains an associated single-guide RNA (sgRNA) molecule, whose first 20 nucleotides are complementary to the DNA target.[3] Thus, Cas9 can generate DSBs at specific locations in genomic DNA when a specific sgRNA is present, providing an unprecedented tool for editing genomes. Once a DSB is created by Cas9 and the specific sgRNA, cellular DNA repair machinery is recruited to fix the DSB. If the error-prone nonhomologous end joining DNA repair system is utilized during this process, mistakes are made and consequently small deletions or insertions are introduced, which lead to the disruption of gene functions.[4–8] In the presence of a repair template, the DSB can be repaired through homologous recombination-mediated DNA repair, which can be effectively used for gene replacement or introduction of new gene functions.[9] Therefore, CRISPR/Cas9 gene editing technology can be used to disrupt gene functions (gene knockout), to replace a mutated gene with a wild-type copy (gene repair), or to introduce a new gene function (gene replacement) such as a fluorescence marker at a desired location in the genome.

In theory, editing a gene only requires the production of both Cas9 protein and a sgRNA molecule that targets the gene of interest and that matches the sequences immediately adjacent to the 5′-end of the NGG protospacer adjacent motif (PAM) site.[1,2] However, production of the CRISPR/Cas9 components in vivo actually can be more nuanced. As discussed later, production of sgRNA molecules can be especially complicated. In addition, the most labor-intensive part of CRISPR/Cas9-mediated genome editing is the analysis and characterization of the edited biological materials. It can be challenging to recover plants that have undergone the desired editing events, but no longer contain genes encoding the CRISPR components. In this chapter, we present a ribozyme-based sgRNA production technology, which provides several advantages over the traditional methods of using RNA polymerase III (Pol III) promoters for sgRNA production.[10,11] Moreover, we provide an effective method to quickly identify *Cas9*-free plants that contain a desired mutation in the target region of the gene of interest.[10]

1. PRODUCTION OF CAS9 PROTEIN IN PLANT CELLS

To achieve efficient gene editing in plants, a high level of Cas9 expression is often needed. More importantly, Cas9 protein needs to be located in the nucleus, where the intended gene editing events take place. Usually a strongly and broadly expressed promoter such as the CaMV 35S promoter or the maize ubiquitin promoter is used to produce high levels of Cas9 protein. Cas9 also is often codon optimized so that maximum expression of the nuclease can be achieved. A nuclear localization signal (NLS) such as SV40 NLS is fused to the C-terminal end of Cas9 to ensure that the expressed Cas9 migrates into and persists within the nucleus. As shown in several recent studies in Arabidopsis, the promoters used to drive *Cas9* expression can have a profound impact on gene editing efficiency in germline cells and, thus, the rate of transmission of edited genes to the next generation.[12–14]

Cas9 can also be engineered to become an RNA-dependent DNA-binding protein lacking its usual nuclease activities (dCas9).[15–18] Mutations of the catalytic residues in the nuclease domains such as D10 and H840 completely abolish the nuclease activities of Cas9. The mutated Cas9 can still bind to specific DNA sequences if a sgRNA molecule is present, providing an outstanding platform upon which to attach enzymes that can modify histones or DNA sequence or chromatin structure at specific locations. For example, when a methyl transferase is fused with dCas9, the fusion protein can cause localized modifications to histones at a genome site specified by a particular sgRNA.[19] Fusions of dCas9 with transcription activator or repressors can activate or repress expression of a specific gene as dictated by the sgRNA bound to the modified dCas9 molecule.[15,17,18] More recently, dCas9 fused with a cytidine deaminase enzyme was shown to have the ability to convert C to T (or G to A) within a window of approximately five nucleotides, providing a very useful tool to introduce point mutations at specific locus defined by a sgRNA molecule.[20] Cytidine deaminase fused with dCas9 has been used to successfully introduce point mutations at specific locations in plants.[21,22]

2. PRODUCTION OF sgRNA MOLECULES IN VIVO

A guide RNA molecule is a central component for CRISPR/Cas9-mediated genome editing and it is the sole determinant for selecting the specific cleavage sites in the genome. A sgRNA can be arbitrarily divided into two parts: the first 20 bp at the 5′-end of a sgRNA complements the target DNA and determines the specificity of the sgRNA (Fig. 1A). We call this part the gene-specificity portion of a sgRNA. The rest of the sgRNA is the core sequence that forms a unique secondary structure to guide the interactions with the Cas9 protein and to bring Cas9 to the DNA target (Fig. 1A). When designing a sgRNA, only the gene-specificity portion needs to be changed. Guide RNA can be chemically synthesized for in vitro studies, but production of a sgRNA in vivo is achieved by transcribing the corresponding DNA sequences (Fig. 1).

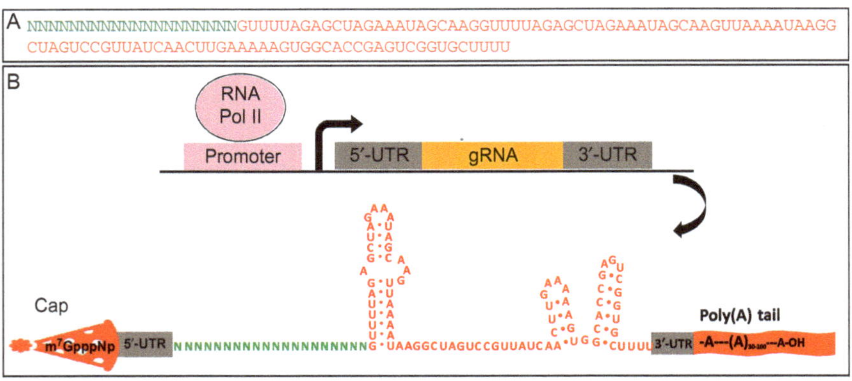

Fig. 1 RNA polymerase II promoters are not suitable for guide RNA production in vivo. (A) A guide RNA molecule for CRISPR/Cas9-mediated gene editing. The sequence N (in *green*) matches the target sequence in the gene of interests. The sequence in *red* is the core sequence of all sgRNAs used in conjunction with Cas9 from *Streptococcus pyogenes*. (B) Transcription of an sgRNA from an RNA polymerase II promoter. The primary transcripts would undergo extensive modifications including 5′-end capping and 3′-end polyadenylation. Such modifications will inhibit the ability of sgRNAs to interact with Cas9 and render sgRNAs nonfunctional. Moreover, the transcripts are actively exported from the nucleus, preventing it from editing DNA.

2.1 RNA Polymerase II Promoters Are Not Suitable for Direct sgRNA Production

Guide RNAs need to meet several requirements in order to edit genes successfully. First, a sgRNA has to be localized in the nucleus. Otherwise, it cannot bring Cas9 to the target DNA to generate DSB, which is a prerequisite for subsequent editing events. Second, both the 5′-end and 3′-end should not be extensively modified. Modifications at either end may interfere with the formation of the RNA:DNA duplex at the target site and/or interactions between sgRNA and Cas9, thus rendering the sgRNA nonfunctional for gene editing.

Such requirements for a functional sgRNA in gene editing essentially eliminate the direct use of any RNA polymerase II (Pol II) promoters for the production of sgRNA in plants or in other organisms (Fig. 1B). The majority of well-characterized promoters are RNA Pol II promoters such as CaMV 35S promoter, ubiquitin gene promoters, and actin gene promoters. RNA Pol II promoters are responsible for the transcription of genes to produce precursors of all mRNA, microRNA, and most of small nuclear RNAs. Primary transcripts from RNA Pol II promoters undergo extensive posttranscriptional processing and modifications including 5′-capping, 3′-polyaddenylation, and removal of introns by splicing (Fig. 1B). Once processed, the majorities of mature RNAs including all mRNAs are transported to cytosol. Therefore, transcripts from RNA Pol II promoters are not suitable for serving as sgRNAs for gene editing (Fig. 1B).

2.2 Guide RNAs Are Mainly Produced From RNA Pol III Promoters

Production of functional sgRNAs in vivo has been achieved almost exclusively through the use of RNA Pol III promoters (Fig. 2A), which direct the biosynthesis of 5S rRNA, tRNA, and some nuclear small RNAs such as U6 snRNA and U3 snRNA. So far, sgRNA production in animal systems has been accomplished exclusively by using a U6 snRNA promoter.[1,2] In plants, both U6 and U3 promoters have been used to express sgRNA molecules for gene editing.[8,23] Guide RNA molecules produced from U6 or U3 promoters have been used successfully in editing various genes in Arabidopsis, rice, and many other organisms.[10,14,22–29]

Despite the success of RNA Pol III promoter-driven sgRNA production in many systems, there are several limitations for U6 promoter or U3 promoter-based sgRNA production. First of all, *U6 snRNA* and *U3 snRNA*

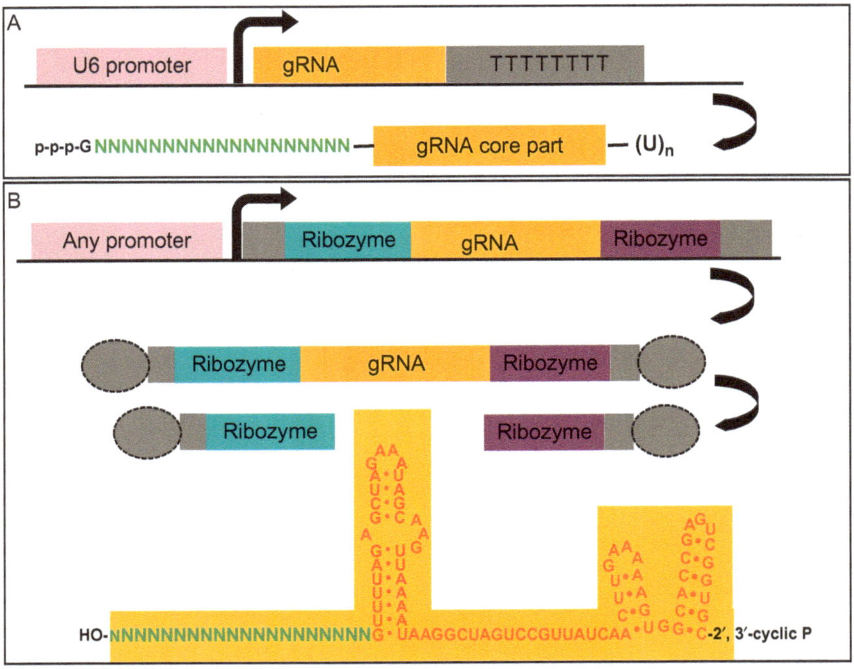

Fig. 2 Production of functional sgRNAs in vivo. (A) Transcription of an sgRNA gene from a U6 promoter. The 5′-end of the transcripts is guanine with a triphosphate group. The 3′-end of the transcripts has a stretch of U with variable length. The first 20 bp sequence (G pluses19 N in *green*) matches the target sequence immediately upstream of the PAM site in the gene of interests. "p–p–p-" refers to triphosphate group at the 5′-end of the sgRNA. $(U)_n$ refers to polyuridine residues at the 3′-end of the sgRNA. (B) Production of a sgRNA from the ribozyme-flanked artificial gene *RGR*. Either RNA polymerase II or RNA Pol III promoters are suitable for guiding the RNA production from RGR. "HO-" refers to a hydroxyl group at the 5′-end of the sgRNA." -2′,3′-Cyclic P" refers to a 2′,3′-cyclic phosphate at the 3′-end of the sgRNA. The sequence N (in *green*) matches the target sequence in the gene of interests. Note that the first base in sgRNA does not have to be a G. The sequence in *red* is the core sequence of all sgRNAs used in conjunction with Cas9 from *Streptococcus pyogenes*.

genes are constitutively and ubiquitously expressed. Therefore, their expression is not tunable and the promoters certainly cannot be used to produce sgRNA in specific cells or tissues. Second, it is not an easy task to choose the most appropriate RNA Pol III promoters because there are usually many copies of U6 or U3 promoters in a genome. Moreover, the majority of U6 and U3 promoters in many organisms have not been characterized. Sometimes, a regulatory element is actually embedded in the coding region of the U6 or U3 snRNA, making it difficult to clone properly configured, active U6 or U3 promoter. Third, transcription from U6 promoters always initiates at a G and U3 promoters begin transcription initially at an A. Therefore, the best CRISPR target sequences are $G(N)_{20}GG$ and $A(N)_{20}GG$, for U6 and U3 promoter-produced sgRNAs, respectively. The restriction on transcription initiation site adds another constrain for target selection. In addition, the U6 and U3 promoters are not suitable for routine in vitro production of sgRNA molecules because RNA Pol III enzyme is not commercially available.

Although primary transcripts from U6 promoters are not as extensively modified as transcripts from RNA Pol II promoters (Fig. 1B), there are still unnecessary extra bases or modifications in U6 transcripts (Fig. 2A). Transcription from U6 promoter terminates at a cluster of thymidines (Fig. 2A). Consequently, U6 snRNA usually contains a uridine stretch of variable length.[30] Currently the most widely used Cas9 in genome engineering is from *Streptococcus pyogenes* (SP) (SpCas9), which uses NGG as its PAM site.[1–3] SpCas9 appears to be able to tolerate the 3′-Poly U and other modifications in sgRNA molecules.[1–3] However, it is not known whether editing efficiency in vivo may actually be improved if the Poly U stretch is removed. It is conceivable that cellular proteins may bind to Poly U and interfere with gene editing. It is likely that Cas9-like molecules from other sources may very well not tolerate the 3′-end uridine stretch at all. Therefore, development of technologies that produce the exact sgRNA molecules without extra bases or modification may greatly benefit the applications using other Cas9-like proteins and other sgRNA-dependent nucleases such as Cpf1 in genome editing.[31,32]

3. PRODUCTION OF sgRNAs FROM RIBOZYME-FLANKED ARTIFICIAL GENE

Because any extra bases or modifications at either end of a sgRNA may render the sgRNA nonfunctional or decrease gene-editing efficiency, ideally sgRNAs should be produced in vivo without such modifications to ensure maximum efficiency. The current methods of sgRNA production also lack cell and tissue specificity. We developed a strategy that produces clean sgRNAs that have been shown to work with temporal and spatial precision (Fig. 2B).[11] We took advantage of ribozymes, which are RNA molecules with enzymatic activities.[33] Some ribozymes have nuclease activities that catalyze the cleavage of an RNA molecule at a specific site(s). The idea is to flank the desired sgRNA molecule with ribozymes at both

5′ and 3′ ends. We designed the ribozyme-flanked sgRNA in such a way that the molecule would undergo self-catalyzed cleavage to release the designed mature sgRNA for gene editing (Fig. 2B).[11] We call this three-way chimera a ribozyme–gRNA–ribozyme unit or, for brevity, an RGR. Such ribozyme-flanked RNA molecules can be easily produced if the DNA encoding an RGR is placed under the control of an appropriate promoter (Fig. 2B). Because both the 5′-side and the 3′-side ribozymes undergo cleavage away from the designed sgRNA because of their intrinsic nuclease activities, any modifications such as capping or polyadenylation will not affect the designed sgRNA (Fig. 2B).[11] Therefore, an RGR can be transcribed from any promoter including RNA Pol II promoters as well as RNA Pol III promoters such as U6 promoters (Fig. 2B).

There are several distinct differences between the sgRNA molecules produced from U6 promoters and sgRNA molecules generated from the ribozyme-based technology. U6 promoter transcribed sgRNA molecules have a triphosphate group at the 5′-end, whereas sgRNA from RGR unit possesses a hydroxyl group at the 5′-end (Fig. 2). The 3′-end of U6-produced sgRNA usually has a stretch of several uridine residues while RGR-produced sgRNAs do not include the poly U (Fig. 2). Moreover, sgRNA produced from RGR possesses a 2′,3′-cyclic phosphate terminus, whereas U6-produced sgRNA usually has a 3′-OH group (Fig. 2A and B).

3.1 Design of RGRs for sgRNA Production

There are several factors that need to be taken into consideration when designing an RGR unit. The cleavage site of the 5′-end ribozyme should be immediately before the first nucleotide of the sgRNA (corresponding to the first nucleotide of the target sequence), and the cleavage site of the 3′-end ribozyme should be immediately after the last nucleotide of the sgRNA. Thus, no extra bases are added to the designed sgRNA. A ribozyme can be used in an RGR unit if (1) it possesses a robust self-cleavage activity, (2) it is small in size, and (3) its activities do not depend on or have very minimum dependency on flanking sequences.

The first 20 bp of guide RNA molecules varies according to different target sequences. A ribozyme whose self-cleavage activity is independent of flanking sequence would be ideal so that universal primers can be used when assembling a new RGR unit. However, all known ribozymes either require certain conserved sequences downstream of the cleavage site or need certain secondary structures around the cleavage site. Hammerhead ribozymes appear to be the best option for 5′-end ribozyme.[33] Hammerhead ribozymes are small and their core sequence are less than 50 bp in length (Fig. 3), allowing chemical synthesis or PCR amplification to generate the corresponding DNA sequences. Only a few invariant bases are required for ribozyme activity as long as the stem-loop structures are maintained. This property makes it easy to assemble multiple RGR units without using the exact same primary sequences.

The 3′-end of a sgRNA molecule is part of the core sequence that does not vary according to target sequences. Therefore, there are more options in choosing a 3′-end ribozyme. Essentially all ribozymes with self-cleavage nuclease activities can be used if the 3′-end of the guide RNA meets the sequence requirement upstream of the cleavage site. A Hammerhead ribozyme may also be used at the 3′-end of an RGR unit. We found that hepatitis delta virus (HDV) ribozymes may be a better choice because they have no requirements for the sequence upstream of the cleavage site.[34] Therefore,

Fig. 3 An example of a complete RGR unit for sgRNA production. The RNA molecule (*top*) was transcribed from a ribozyme-flanked artificial gene *RGR*. The sequence in *black* and the *underlined* sequence assemble into a Hammerhead ribozyme. Sequence in *purple* targets the gene of interests. Note that part of the target sequence is included in the Hammerhead ribozyme. The first 6 bp of Hammerhead must be reverse complementary to the first 6 bp of the sgRNA. The sequence in *red* is the core sequence of all sgRNAs. The sequence in *green* is a HDV ribozyme. The RNA molecule is self-folded and processed into ribozymes and the mature sgRNA (*bottom*). The scissor in *blue* shows the self-cleavage site of the Hammerhead ribozyme. The scissor in *green* shows the self-processing site of the HDV ribozyme.

the same universal sequence could be used for any guide RGR constructs (Fig. 3).

An example of a complete RGR is shown in Fig. 3. This RGR can be placed under the control of any promoter for sgRNA production. The primary transcripts undergo self-cleavage to release the mature sgRNA as predicated. This particular RGR was transcribed from a Sp6 promoter in vitro.[11] The primary transcripts underwent the expected self-cleavage in vitro to release the mature sgRNA, which, in turn, was demonstrated to bind Cas9 and guide the nuclease to a specific target DNA where it catalyzed a DSB.[11] It is worth mentioning that there was no need to purify the mature sgRNA from the in vitro transcription products for Cas9-mediated DNA cleavage in vitro. The transcription reaction mixture that contains the ribozymes, mature sgRNA, and other chemicals was satisfactory for in vitro Cas9-mediated DNA digestion.[11] When the RGR was placed under the control of the *Alcohol Dehydrogenase 1* gene promoter (*ADH1* promoter), which is an RNA Pol II promoter, functional sgRNA molecules were released from the transcripts and consequently the target gene was edited as designed when Cas9 was provided.[11]

The ribozyme-based technology has been used successfully to edit genes in plants.[10,35] For example, RGR-based production of a sgRNA targeting the *Auxin Binding Protein 1* (*ABP1*) gene, which encodes a protein that binds auxin and that has been claimed as an auxin receptor, led to the identification of a series of loss-of-function *abp1* mutants.[35] The new *abp1* mutants displayed no obvious defects in development and auxin response, providing unambiguous evidence that ABP1 is not required for either auxin signaling or Arabidopsis development.

3.2 Advantages of the Ribozyme-Based Technology

Ribozyme-based sgRNA production opens a door for conducting more sophisticated gene editing experiments that would not be possible by using RNA Pol III promoter-based technology. Because primary transcripts from RGR are automatically processed to release the sgRNA molecule, we can use any promoters for the production of sgRNA (Fig. 2B). Cell/tissue specific promoters would enable gene editing in specific cells. This is especially important for gene therapy or other biomedical applications. Guide RNAs can also be produced using inducible promoters or promoters regulated by environmental signals such as light greatly expanding the scope of gene editing in basic research as well for applications in crop improvement and development of new treatments in biomedicine. Production of sgRNA with temporal/spatial control is especially important when we use dCas9 to activate or repress gene expression, enabling genes to be temporally and spatially regulated.

The RGR-based sgRNA production can be further coupled with the production of a fluorescence marker protein.[36] For example, the *mKate2*, which encodes a red fluorescence protein, can be placed upstream of an RGR gene with an RNA triplex in between.[36] The entire unit can be transcribed from a single promoter. The primary transcripts will undergo self-processing to release *mKate2* mRNA, the mature sgRNA, and the ribozymes. The *mKate2* mRNA is transported into cytosol to direct the synthesis of the fluorescence protein. In combining both RGR and a fluorescence maker, not only is a functional sgRNA produced, but also the cells that undergo gene editing are marked. It is conceivable that this technology coupled with cell sorting will open completely new lines of research. For example, when cell-specific gene editing is performed, we can profile gene expression, hormone levels, and metabolic changes in the edited cells.

RGRs can be easily made in tandem, providing a facile way to multiplex gene editing. One big advantage of the ribozyme-based technology is that each sgRNA concentration will be similar because they are transcribed from the same promoter. Another advantage is that the ribozyme sequences can be varied as long as the secondary structures are maintained, which decreases the number of exact repeat sequences and which makes assembly of tandem RGRs a less laborious task.

4. EFFICIENT IDENTIFICATION OF *CAS9*-FREE PLANTS WITH THE DESIRED EDITING EVENTS

Successful isolation of stable and heritable mutants generated by CRISPR/Cas9 relies on two factors: (1) The editing efficiency is high enough; (2) An effective selection method that easily differentiates wild-type plants from plants with heritable mutations. Most efforts in plant gene editing have focused on improving editing efficiency.[12–14] For example, strong promoters such as *Yao* promoters or reproductive tissue-specific promoters including *EC2* promoter were used to drive the expression of the Cas9 protein.[13,14] Use of such promoters has resulted in a great increase in editing efficiency compared to CaMV 35S promoter-mediated Cas9 expression. We focused on improving the efficiency of screening for useful edited products in Arabidopsis.[10] The concept can also be easily adapted for other plants. Mutations generated by CRISPR/Cas9 in Arabidopsis are usually somatic and very few mutations are transmitted to next generations.[10] Sometimes it is very misleading that mutations are detected in T1 and T2 plants, giving a false sense that heritable mutations have been obtained. However, once the T2 plants are crossed to another background, the previously detected mutations often disappeared.[10]

It is clear that Cas9/sgRNA will continue to generate mutations if the target sequences are still intact. However, it is difficult to differentiate between the mutations inherited from the previous generation and new mutations generated by CRISPR/Cas9 in the examined generation. Therefore, we believe that isolation of mutations in *Cas9*-free Arabidopsis plants is very important, which can ensure that the identified mutations are indeed transmitted from the previous generation and which can be reliably transmitted to next generations. Traditionally, a PCR-based method is used to screen for *Cas9*-free plants, but such methods are laborious because it takes time and labor to grow plants, extract DNA, and conduct PCR analysis. We added a cassette that expresses a fluorescence marker, mCherry, in Arabidopsis to the CRISPR/Cas9 construct (Fig. 4).[10] Therefore, we can use mCherry fluorescence as a proxy for the presence of the CRISPR/Cas9 construct. As a result, *Cas9*-free plants can be easily identified at the seed stage, eliminating the laborious work of planting many *Cas9*-containing plants.

The method worked very efficiently in Arabidopsis as we expected (Fig. 4). We used this method to generate mutations in the *ABP1* gene in Arabidopsis as a proof of concept.[10] We found that less than 50% of T1 plants, which contained detectable mutations, produced T2 offspring that are *Cas9*-free and that contain the expected deletions/insertions. Moreover, we found that the chance for identifying a mutation in *Cas9*-free Arabidopsis T2 plants is actually very low. For example, we genotyped 95 *Cas9*-free T2 plants generated from a single T1 plant and only one out of the 95 plants contained a mutation in the *ABP1* gene.[10] If the T2 plants were not first screened for *Cas9*-free plants, much more effort would be needed to identify a reliable and heritable mutation. In this particular case, we would have to genotype at least 75% more T2 plants (assuming that only one T-DNA insertion in the T1 plant) if we did not preselect the *Cas9*-free plants. Because there are usually at least two T-DNA insertions in a T1 plant, we would have to genotype more than 380 T2 plants in this case to identify one Cas9-free plant with a heritable mutation. The workload for isolating a heritable and stable mutation in Arabidopsis would become prohibitive if we would not be able to

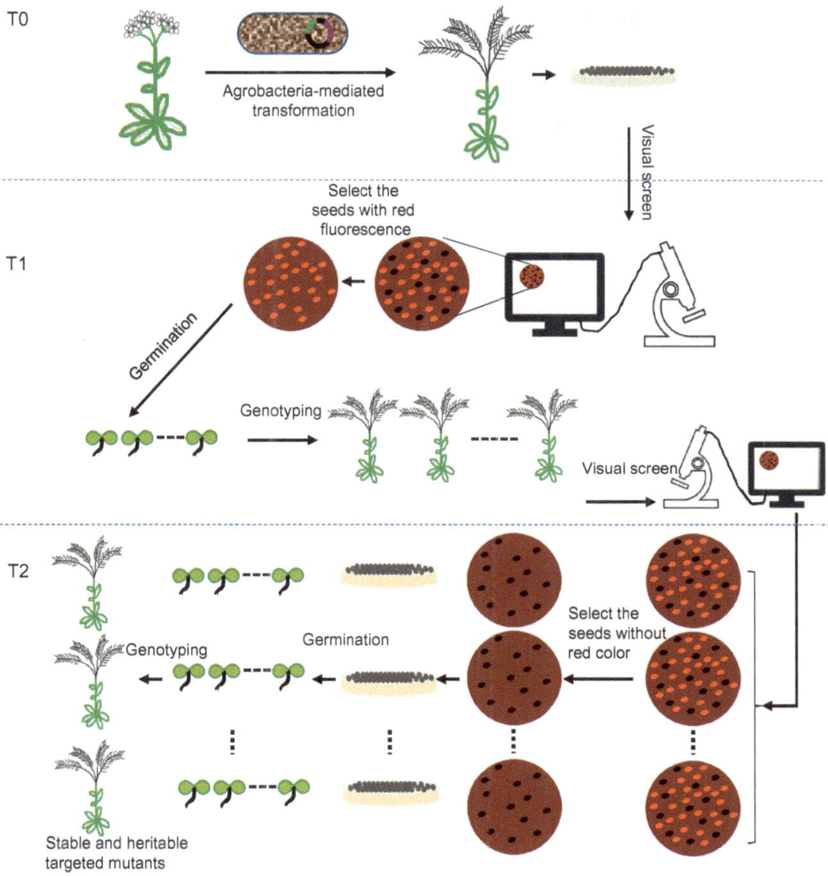

Fig. 4 A fluorescence-based method for reliably identifying stable and heritable mutations generated by CRISPR/Cas9 in Arabidopsis. At T0 generation, a CRISPR/cas9 construct with the mCherry expression unit is delivered into Arabidopsis through Agrobacteria-mediated transformation. Transgenic plants are identified either by antibiotic-resistance or by mCherry fluorescence at T1 generation. T1 plants are analyzed by PCR/restriction digestion to identify putative mutants. In T2 generation, Cas9-free T2 seeds (no red fluorescence) are identified by mCherry fluorescence with a fluorescent microscope. Mutations identified at T2 generation in the Cas9-free plants are transmitted from the previous generation and are stable and heritable.

preselect the *Cas9*-free T2 plants because not every T1 plants with a detectable mutation can transmit the detected mutations to the next generation. Inclusion of the *mCherry* expression cassette in the CRISPR construct makes the selection of *Cas9*-free T2 seeds very convenient and efficient, enabling us to reliably identify heritable and stable mutations in Arabidopsis.

In summary, we provide a powerful method to produce sgRNA molecules in vitro and in vivo. The ribozyme-based sgRNA production offers many advantages over the traditional U6-based methods. With ever expanding genome editing nucleases, the RGR method will see an even more predominant role in gene editing because the ribozyme-based technology produces sgRNA molecules with more accurate ends without extra bases. The mCherry-based method provides an efficient and reliable approach to quickly identify transgene-free plants whose genomes have been properly edited.

ACKNOWLEDGMENT

This work is supported by a grant from National Institutes of Health (GM114660 to Y.Z.) and a startup from the Huazhong Agricultural University.

REFERENCES

1. Cong L, Ran FA, Cox D, et al. Multiplex genome engineering using CRISPR/Cas systems. *Science*. 2013;339(6121):819–823.
2. Mali P, Yang L, Esvelt KM, et al. RNA-guided human genome engineering via Cas9. *Science*. 2013;339(6121):823–826.
3. Jinek M, Chylinski K, Fonfara I, Hauer M, Doudna JA, Charpentier E. A programmable dual-RNA-guided DNA endonuclease in adaptive bacterial immunity. *Science*. 2012;337(6096):816–821.
4. Ma X, Zhu Q, Chen Y, Liu YG. CRISPR/Cas9 platforms for genome editing in plants: developments and applications. *Mol Plant*. 2016;9(7):961–974.
5. Char SN, Neelakandan AK, Nahampun H, et al. An agrobacterium-delivered CRISPR/Cas9 system for high-frequency targeted mutagenesis in maize. *Plant Biotechnol J*. 2016;15:257–268.
6. Xie K, Minkenberg B, Yang Y. Boosting CRISPR/Cas9 multiplex editing capability with the endogenous tRNA-processing system. *Proc Natl Acad Sci USA*. 2015; 112(11):3570–3575.
7. Lowder LG, Zhang D, Baltes NJ, et al. A CRISPR/Cas9 toolbox for multiplexed plant genome editing and transcriptional regulation. *Plant Physiol*. 2015;169(2):971–985.
8. Miao J, Guo D, Zhang J, et al. Targeted mutagenesis in rice using CRISPR-Cas system. *Cell Res*. 2013;23(10):1233–1236.
9. Sun Y, Zhang X, Wu C, et al. Engineering herbicide-resistant rice plants through CRISPR/Cas9-mediated homologous recombination of acetolactate synthase. *Mol Plant*. 2016;9(4):628–631.
10. Gao X, Chen J, Dai X, Zhang D, Zhao Y. An effective strategy for reliably isolating heritable and Cas9-free Arabidopsis mutants generated by CRISPR/Cas9-mediated genome editing. *Plant Physiol*. 2016;171(3):1794–1800.
11. Gao Y, Zhao Y. Self-processing of ribozyme-flanked RNAs into guide RNAs in vitro and in vivo for CRISPR-mediated genome editing. *J Integr Plant Biol*. 2014; 56(4):343–349.
12. Mao Y, Zhang Z, Feng Z, et al. Development of germ-line-specific CRISPR-Cas9 systems to improve the production of heritable gene modifications in Arabidopsis. *Plant Biotechnol J*. 2016;14(2):519–532.

13. Yan L, Wei S, Wu Y, et al. High-efficiency genome editing in Arabidopsis using YAO promoter-driven CRISPR/Cas9 system. *Mol Plant*. 2015;8(12):1820–1823.
14. Wang ZP, Xing HL, Dong L, et al. Egg cell-specific promoter-controlled CRISPR/Cas9 efficiently generates homozygous mutants for multiple target genes in Arabidopsis in a single generation. *Genome Biol*. 2015;16:144.
15. Komor AC, Badran AH, Liu DR. CRISPR-based technologies for the manipulation of eukaryotic genomes. *Cell*. 2016;168:20–36.
16. Gao Y, Xiong X, Wong S, Charles EJ, Lim WA, Qi LS. Complex transcriptional modulation with orthogonal and inducible dCas9 regulators. *Nat Methods*. 2016;13(12):1043–1049.
17. Dominguez AA, Lim WA, Qi LS. Beyond editing: repurposing CRISPR-Cas9 for precision genome regulation and interrogation. *Nat Rev Mol Cell Biol*. 2016; 17(1):5–15.
18. Gilbert LA, Larson MH, Morsut L, et al. CRISPR-mediated modular RNA-guided regulation of transcription in eukaryotes. *Cell*. 2013;154(2):442–451.
19. Liu XS, Wu H, Ji X, et al. Editing DNA methylation in the mammalian genome. *Cell*. 2016;167(1):233–247:e217.
20. Komor AC, Kim YB, Packer MS, Zuris JA, Liu DR. Programmable editing of a target base in genomic DNA without double-stranded DNA cleavage. *Nature*. 2016; 533(7603):420–424.
21. Lu Y, Zhu JK. Precise editing of a target base in the rice genome using a modified CRISPR/Cas9 system. *Mol Plant*. 2016;10:523–525.
22. Li J, Sun Y, Du J, Zhao Y, Xia L. Generation of targeted point mutations in rice by a modified CRISPR/Cas9 system. *Mol Plant*. 2016;10:526–529.
23. Li JF, Norville JE, Aach J, et al. Multiplex and homologous recombination-mediated genome editing in Arabidopsis and Nicotiana benthamiana using guide RNA and Cas9. *Nat Biotechnol*. 2013;31(8):688–691.
24. Woo JW, Kim J, Kwon SI, et al. DNA-free genome editing in plants with preassembled CRISPR-Cas9 ribonucleoproteins. *Nat Biotechnol*. 2015;33(11):1162–1164.
25. Mikami M, Toki S, Endo M. Comparison of CRISPR/Cas9 expression constructs for efficient targeted mutagenesis in rice. *Plant Mol Biol*. 2015;88(6):561–572.
26. Zhang H, Zhang J, Wei P, et al. The CRISPR/Cas9 system produces specific and homozygous targeted gene editing in rice in one generation. *Plant Biotechnol J*. 2014;12(6):797–807.
27. Shan Q, Wang Y, Li J, Gao C. Genome editing in rice and wheat using the CRISPR/Cas system. *Nat Protoc*. 2014;9(10):2395–2410.
28. Schiml S, Fauser F, Puchta H. The CRISPR/Cas system can be used as nuclease for in planta gene targeting and as paired nickases for directed mutagenesis in Arabidopsis resulting in heritable progeny. *Plant J*. 2014;80(6):1139–1150.
29. Feng Z, Mao Y, Xu N, et al. Multigeneration analysis reveals the inheritance, specificity, and patterns of CRISPR/Cas-induced gene modifications in Arabidopsis. *Proc Natl Acad Sci USA*. 2014;111(12):4632–4637.
30. Licht K, Medenbach J, Luhrmann R, Kambach C, Bindereif A. 3′-Cyclic phosphorylation of U6 snRNA leads to recruitment of recycling factor p110 through LSm proteins. *RNA*. 2008;14(8):1532–1538.
31. Zetsche B, Heidenreich M, Mohanraju P, et al. Multiplex gene editing by CRISPR-Cpf1 using a single crRNA array. *Nat Biotechnol*. 2016;35:31–34.
32. Fonfara I, Richter H, Bratovic M, Le Rhun A, Charpentier E. The CRISPR-associated DNA-cleaving enzyme Cpf1 also processes precursor CRISPR RNA. *Nature*. 2016;532(7600):517–521.
33. Doudna JA. Ribozymes: the hammerhead swings into action. *Curr Biol*. 1998;8(14): R495–R497.

34. Thill G, Blumenfeld M, Lescure F, Vasseur M. Self-cleavage of a 71 nucleotide-long ribozyme derived from hepatitis delta virus genomic RNA. *Nucleic Acids Res.* 1991;19(23):6519–6525.
35. Gao Y, Zhang Y, Zhang D, Dai X, Estelle M, Zhao Y. Auxin binding protein 1 (ABP1) is not required for either auxin signaling or Arabidopsis development. *Proc Natl Acad Sci USA*. 2015;112(7):2275–2280.
36. Nissim L, Perli SD, Fridkin A, Perez-Pinera P, Lu TK. Multiplexed and programmable regulation of gene networks with an integrated RNA and CRISPR/Cas toolkit in human cells. *Mol Cell.* 2014;54(4):698–710.

CHAPTER TEN

Engineering Molecular Immunity Against Plant Viruses

Syed Shan-e-Ali Zaidi*,†, Manal Tashkandi*, Magdy M. Mahfouz*,[1]

*Laboratory for Genome Engineering, 4700 King Abdullah University of Science and Technology, Thuwal, Saudi Arabia

†National Institute for Biotechnology and Genetic Engineering (NIBGE), Faisalabad, Pakistan

[1]Corresponding author: e-mail address: magdy.mahfouz@kaust.edu.sa

Contents

Abstract

Genomic engineering has been used to precisely alter eukaryotic genomes at the single-base level for targeted gene editing, replacement, fusion, and mutagenesis, and plant viruses such as *Tobacco rattle virus* have been developed into efficient vectors for delivering genome-engineering reagents. In addition to altering the host genome, these methods can target pathogens to engineer molecular immunity. Indeed, recent studies have shown that clustered regularly interspaced short palindromic repeats (CRISPR)/CRISPR-associated 9 (Cas9) systems that target the genomes of DNA viruses can interfere with viral activity and limit viral symptoms in planta, demonstrating the utility of this system for engineering molecular immunity in plants. CRISPR/Cas9 can efficiently target single and multiple viral infections and confer plant immunity. Here, we discuss the use of site-specific nucleases to engineer molecular immunity against DNA and RNA viruses in plants. We also explore how to address the potential challenges encountered when producing plants with engineered resistance to single and mixed viral infections.

Progress in Molecular Biology and Translational Science, Volume 149
ISSN 1877-1173
http://dx.doi.org/10.1016/bs.pmbts.2017.03.009

1. INTRODUCTION

The ability to alter DNA sequences in a site-specific manner and to generate gene mutants and protein variants can facilitate the interrogation of gene function and understanding of functional biology. The term "genome engineering" refers to the precise, site-specific manipulation of the genome at the single-base level.[1] Genome engineering uses site-specific nucleases (SSNs) to generate site-specific double-stranded breaks (DSBs) in the genome. DSBs trigger the cellular repair machinery[2] and two major DNA repair pathways, the imprecise nonhomologous end joining (NHEJ) pathway and the precise homology-directed repair (HDR) pathway, can repair DSBs[3] (Fig. 1). NHEJ repair is suitable for site-specific mutagenesis and for subsequent functional knockout of genes. Therefore, genome engineering via NHEJ is ideal for interrogating gene functions. By contrast, HDR uses a DNA template to provide the information needed by the HDR repair machinery to copy the sequence across the break.[4] HDR constitutes a precise form of repair and is ideal for generating gene and protein variants based on a DNA template that is supplied with the nuclease that produces the DSBs. Several genome-engineering platforms have been developed and repeatedly improved, but limitations to their efficiency and ease of use remain, especially in plants.[5] One limitation is the choice of vector for delivering genome-engineering reagents into the host cell.

Work in a variety of systems has used viruses to deliver genome-engineering reagents into target cells.[6] For example, in mammalian systems, the effective delivery of genome-engineering reagents has been achieved using viruses, including retroviruses,[7,8] adenoviruses,[9] and adeno-associated viruses.[10] These viral vectors can be delivered into virtually any human cell type[8,10,11] and function as templates not only for expression of the designer SSNs but also for targeted chromosomal integration of foreign DNA.[6,11] Recent studies have developed DNA and RNA viruses for genome engineering in plants.[12,13]

In addition to using viruses as delivery vectors, genome-engineering reagents can now be used to target invading DNA viruses for degradation and to establish plant molecular immunity against single and multiple viral infections.[14,15] For example, Ali *et al.* engineered tobacco plants with clustered regularly interspaced short palindromic repeats (CRISPR)/CRISPR-associated 9 (Cas9) targeting the *Tomato yellow leaf curl virus* (TYLCV),

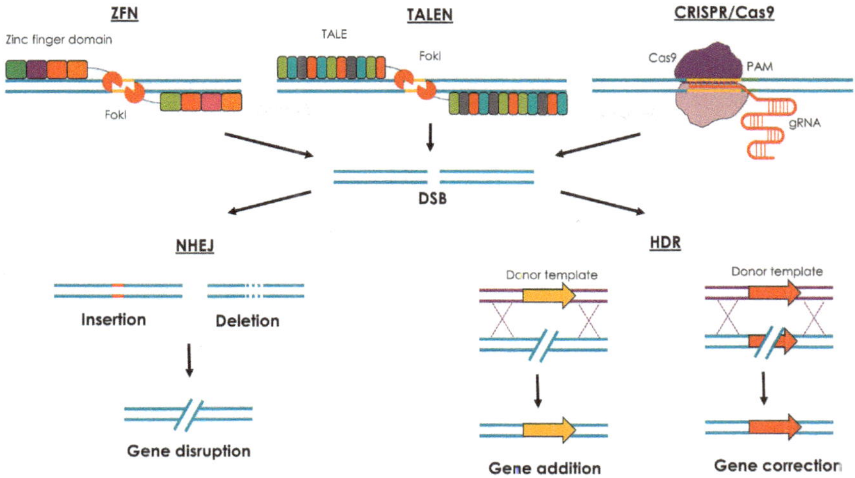

Fig. 1 Genome-engineering platforms and their utilization. The zinc finger nuclease (ZFN) system is composed of a zinc finger domain and a FokI endonuclease domain. The trinucleotide binding sites within each zinc finger domain dictate the specificity of DNA binding, while coupling of the nuclease domains (derived from the type II restriction endonuclease FokI) of two ZFNs produces a double-stranded break (DSB) at the designated target site. Transcription activator-like effector nucleases (TALENs) also contains two domains, a DNA binding domain and a FokI endonuclease domain. But in this case, the DNA binding specificity is dictated by the TALE array that contains 33–34 nucleotide conserved residue repetitive motifs. These motifs contain repeated variable diresidues (RVDs) at positions12 and 13 that specify binding to either an A, G, C, or T nucleotide in the DNA target site. Similar to ZFN, the dimerized FokI endonuclease produces a DSB at the target site. The clustered regularly interspaced short palindromic repeat (CRISPR)/CRISPR-associated 9 (Cas9) system consists of Cas9 endonuclease and a single-guide RNA (sgRNA). In case of CRISPR/Cas9, the DNA binding specificity is dictated by specific hybridization to the sgRNA where the two Cas9 nuclease domains produce a DSB three base pairs upstream of the protospacer adjacent motif (PAM). The repair of the DSB is either by the error-prone nonhomologous end joining (NHEJ) system or by homology-directed repair (HDR). During NHEJ, random insertions or deletions can occur, which ultimately may lead to gene disruption. By contrast, HDR uses a donor DNA template that can be used to direct gene replacement/addition or gene sequence modification.

establishing virus interference and significantly limiting viral symptoms.[14] Similarly, Baltes *et al.* and Ji *et al.*[16,17] established virus interference in tobacco against *Bean yellow dwarf virus* (BeYDV) and in *Arabidopsis thaliana* against *Beet severe curly top virus* (BSCTV). Here, we discuss the current research on the use of genome engineering to develop virus interference strategies and address the potential challenges of engineering plants with effective, durable virus resistance.

2. THE USE OF SSNs FOR GENOME ENGINEERING

SSNs, including zinc finger nucleases (ZFNs), transcription activator-like effector nucleases (TALENs),[18–23] and the recently developed CRISPR/Cas9[24–26] system, have been used to produce targeted mutations and edit the genomes of various eukaryotic species[1] (Fig. 1). ZFNs and TALENs have been used for targeted editing of plant genomes,[27–29] but the specificity of these SSNs[30] comes from protein–DNA binding and therefore ZFNs and TALENs require protein engineering for each user-selected target sequence.[31] Engineering ZFNs and TALENs are time-consuming and resource-intensive and must be performed by skilled molecular biologists.[32,33] Several repeat assembly protocols have been developed for TALEN engineering, but the need for two TALEN monomers to bind to the sense and antisense strands simultaneously complicates the broad utility and applicability of this technique in plants.[4] These limitations have been largely overcome in the recently developed CRISPR/Cas9 system.[34–36] CRISPR/Cas9 engineering is highly predictable and much easier to perform than the other techniques, with greater reproducibility and affordability.[37]

In nature, the CRISPR/Cas9 system provides adaptive molecular immunity to bacterial and archaeal species to confer resistance against invading DNA species, including conjugative plasmids and phages.[38] The CRISPR/Cas9 locus in the bacterial genome is typically composed of Cas genes and the repeat and spacer sequences downstream of a leader sequence.[39,40] CRISPR/Cas9-based immunity is achieved via three distinct phases: foreign DNA is acquired from the plasmid or phage (acquisition phase), the CRISPR locus is transcribed and CRISPR RNA processing and maturation take place (transcription and maturation phase), and the paired transactivating crRNAs (tracrRNAs) and CRISPR RNAs (crRNAs) are used as guides for the Cas9 endonuclease to target or interfere with subsequent invasions (interference phase).[39,41,42] Therefore, the CRISPR/Cas9 system acquires pieces of DNA from an invading phage or plasmid and uses these sequences as molecular records to guide Cas9 endonuclease to bind to the phage or plasmid DNA in subsequent invasions, binding in a region preceding the protospacer adjacent motif (PAM) sequence.[43] The PAM, a trinucleotide sequence that is important for Cas9 activity, is used to distinguish self from non-self-DNA.[44] The PAM sequence of *Streptococcus pyogenes* Cas9 is NGG (with N being any nucleotide) and that of *Scheloribates thermophilus* Cas9 is NGGNG.[43,45]

Notably, the tracrRNAs and crRNAs required for Cas9 specificity can be fused to generate a chimeric single-guide RNA molecule (sgRNA) that can direct Cas9 to a specific sequence preceding the PAM trinucleotide sequence.[46] Since engineering via the CRISPR/Cas9 system involves changing only the 20-nucleotide sequence, the CRISPR/Cas9 system is the system of choice for genome-engineering applications.[26]

The CRISPR/Cas9 system is based on Watson–Crick base pairing between the sgRNA and the DNA of the target genome.[47] Therefore, the CRISPR/Cas9 system supersedes and outperforms other systems due to its ease of use, reproducibility, and affordability.[26,48] The CRISPR/Cas9 system has recently been used for various genome-engineering applications across eukaryotic species, including plants.[49–51] For example, the CRISPR/Cas9 system was used to edit the wheat and maize genomes for targeted trait development.[52–54] Various CRISPR/Cas9-based genome-engineering modalities have been used involving different Cas9 variants[55] or catalytically inactive Cas9 (dCas9) to modulate gene expression patterns and to subsequently interrogate gene functions.[56,57]

Editing of genes and genomes usually involves site-specific mutagenesis, including DNA addition, deletion, or creation of gene fusions. For any genome-engineering platform or modality, there are always concerns about specificity, delivery of reagents, and HDR efficiency in the target cells. Therefore, ongoing, active research aims to improve the specificity, delivery, and editing efficiency in a target cell or species to maximize the usefulness of the genome-engineering modality.[38] Very recently, a novel class of DNA-guided SSNs involving FEN:FokI has been reported; these are expected to be suitable for various engineering purposes.[58]

3. VIRUSES FOR DELIVERY OF GENOME-ENGINEERING REAGENTS

As mentioned in Section 1, many studies in mammalian systems have successfully used viruses to deliver genome-engineering reagents. For plants, the TRV RNA virus has great potential for use as a vector for genome engineering. Important features of the TRV-based delivery system include the following: (1) TRV can systemically infect many plant species, both naturally and under laboratory conditions; (2) TRV can easily be introduced into plants via *Agrobacterium* and delivered into the growing points of the plant; (3) the small genome size of TRV facilitates cloning, multiplexing, library construction, and agroinfection; and (4) the RNA genome of this virus does not integrate into the plant genome. A TRV-based expression system has been developed for transient delivery of ZFNs into a variety of tissues and cells of intact plants, representing a nontransgenic approach for ZFN delivery and mutant plant production.[59] TRV has a bipartite genome, with two RNA components: RNA1 (or TRV1) and RNA2 (or TRV2). In the above-mentioned system, RNA2 was replaced with RNA for the Zif268: FokI ZFN and targeted genome modifications were recovered in somatic tobacco (*Nicotiana tabacum*) and petunia (*Petunia hybrida*) cells at an integrated reporter gene.[59] TRV systemically infected its hosts, and ZFN-mediated, precise mutagenesis was clearly observed in newly developed infected tissues, as measured by the activation of a mutated reporter transgene in tobacco and petunia plants.[59]

If TRV could infect germ cells, this would improve its utility as a vector for genome engineering, making it possible to harvest mutant seeds from infected plants. Indeed, the ability of TRV to move to developing buds and regenerating tissues has enabled researchers to recover mutated plants.[59] Sequence analysis and examination of transmission of the mutations to the next generation confirmed the stability of the ZFN-induced genetic changes.[59] Because TRV can infect a wide range of plant species, it provides a viable alternative to the production of ZFN-mediated mutants while avoiding the use of classical plant transformation methods.

Recently, we developed a TRV system that delivers single and multiple sgRNAs for targeted editing of the host genome and for targeting invading DNA viruses[12,14,15,60] (Fig. 3). In this system, TRV delivers single or multiple sgRNAs for genomic targets to modify the genomes of host plants, including tobacco and *Arabidopsis*. To determine whether the newly developed TRV could produce and deliver sgRNA molecules systemically and to evaluate the persistence of the genomic modification in growing tissues, the engineered TRV was delivered by agroinfiltration and reconstituted in the leaves of *N. benthamiana* expressing Cas9.[12,60] Interestingly, TRV-mediated delivery of sgRNA was not limited to the original infiltrated plants; this precise targeting can also be transmitted to the next generation (Fig. 2). This TRV-mediated sgRNA delivery system stands out because it bypasses the requirement for plant transformation and tissue culture.[12] This technique opens up new possibilities for producing plants with desired traits using CRISPR/Cas9 without the need for laborious, time-consuming intermediate steps.[12]

One potential limit of RNA viruses is that these viruses are limited in the amount of heterologous DNA they can carry (approximately 2–3 kb);

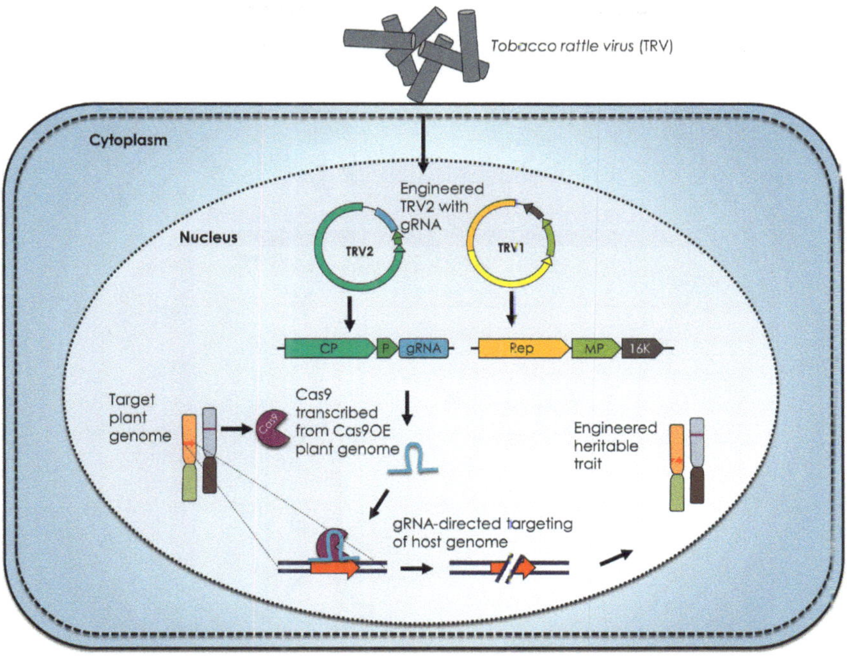

Fig. 2 Molecular basis of CRISPR/Cas9-mediated genome engineering using the TRV system. The engineered TRV genome is transcribed inside the plant cell's nucleus, producing a single guide RNA (sgRNA) under the control of a user-defined promoter. The target plant must be genetically engineered with a Cas9 gene to constitutively produce the Cas9 endonuclease. Cas9 binds the TRV-encoded sgRNA and cleaves the plant genome at the gRNA-directed site (*red arrow*), three base pairs upstream of the PAM sequence. The resulting double-stranded break (DSB) is repaired by the cell's error-prone nonhomologous end joining (NHEJ) DNA repair system. As a result, certain insertions or deletions (indels) are produced, ultimately disrupting the target gene (in most cases). This engineered mutation is a permanent part of the plant's genome and any resulting altered phenotype is transmissible to the next generation.

excess cargo results in the loss of systemic movement or recombinational loss of the cargo DNA.[61,62] This limits the utility of these viruses for delivering TALENs or Cas9 expression cassettes, which typically exceed the virus's cargo capacity. However, for relatively small molecules such as ZFNs, meganucleases, and sgRNAs, TRV was proven to be an effective delivery agent, achieving mutagenesis in somatic and germline cells of several plant species, including tobacco,[59] *N. benthamiana*,[12,60] and petunia.[59] In addition to RNA plant viruses, certain DNA viruses have been used to deliver genome-engineering reagents, including SSNs and DNA templates, for efficient genome editing; readers are directed to the following

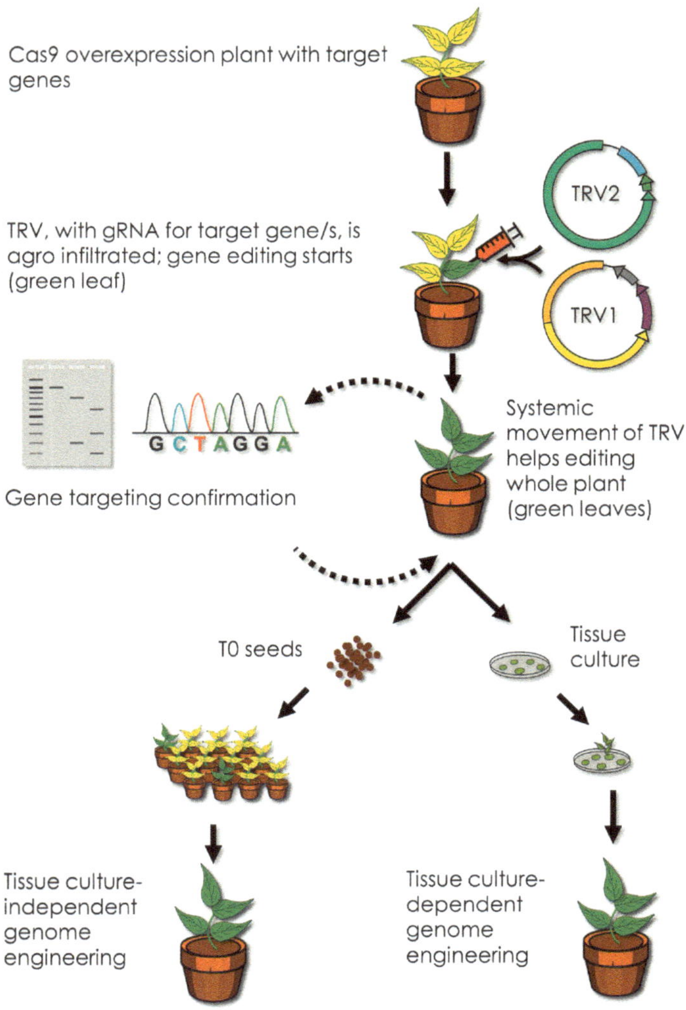

Fig. 3 Stepwise methodology for TRV-mediated genome engineering. In the TRV2 genome, a 20-nucleotide target sequence preceding the PAM sequence is cloned into the sgRNA gene that is under the control of an appropriate promoter. *Agrobacterium* cultures carrying DNA segments encoding the engineered TRV2 genome (conferring user-selected sequence specificity) and the TRV1 genome are coinfiltrated into the leaves of plants overexpressing Cas9 (Cas9-OE) via agroinfection. For example, the sgRNA (cloned in TRV2) is designed to target a malfunctioning gene makes plant leaves yellow. After agroinfection, the plants are analyzed for the presence of the targeted modification that restors the green leaf phenotype. Plant leaf discs carrying modified genomes can be regenerated to recover mutant plants, or the seed progeny can be screened for the presence of the modification in the next generation, thereby bypassing the need for tissue culture.

articles on the development of DNA viruses as vehicles for genome engineering.[13,63–66]

4. USING SSNs TO DEVELOP VIRUS RESISTANCE

As noted above, several studies have reported engineering of SSNs to target viral genomes for degradation. Successful attempts used ZFNs to target TYLCV in *Arabidopsis*[67,68]; resistance developed in transgenic plants expressing ZFNs targeting the TYLCV genome. Initial studies aimed at targeting plant viruses via genome engineering have focused on using ZFN to target begomovirus Rep binding sites/iterons, the iterated nucleotide sequences within the begomovirus genome that provides binding site for virus replication-associated protein (Rep). ZFNs developed for iterons of *Beet severe curly top virus* (BSCTV) and *Tomato yellow leaf curl virus* (TYLCV) effectively targeted the respective viruses.[68,69] Transgenic *Arabidopsis* plants expressing ZFNs that target iterons of BSCTV demonstrated complete resistance to this virus.[69] Chen *et al.* demonstrated that ZFN can efficiently target BSCTV and TYLCV but argued that this viral sequence-specific strategy may not be effective in the field, where mixed viral infections are common.[67] As an alternative, targeting the conserved regions of viruses would confer relatively durable resistance. Three conserved regions in Rep of monopartite begomoviruses were identified and tested for broad-spectrum resistance. Among these three targets, one targeting a 25-base pair sequence functioned well against both *Tomato yellow leaf curl China virus* (TYLCCNV) and *Tobacco curly shoot virus* (TbCSV).[67]

Artificial TALE (ATALE), an artificial DNA-binding protein that has a highly conserved central region with tandem repeat units of 34 amino acids, also has been used for designing broad-spectrum resistance to begomoviruses.[70] Two highly conserved targets, AC1 and a nonanucleotide sequence, were selected and targeted using an ATALE. The DNA binding efficiency of the ATALE was confirmed, and transgenic *N. benthamiana* plants were developed. The authors tested two begomoviruses, *Tobacco curly shoot virus* (TbCSV) and *Tomato leaf curl Yunnan virus* (TLCYnV), as well as one begomovirus with its associated betasatellite, *Tomato yellow leaf curl China virus* (TYLCCNV) with its cognate Tomato yellow leaf curl China betasatellite (TYLCCNB). The resulting transgenic plants demonstrated partial resistance, with delayed symptom development and reduced viral DNA accumulation compared to wild type.[70]

5. CRISPR/Cas9-MEDIATED VIRUS INTERFERENCE

Since the native CRISPR/Cas9 system can confer resistance against invading phages in bacterial and archaeal species, we investigated whether this system could be transferred to plant species to confer molecular immunity against plant DNA viruses just as ZFNs and TALENs were shown to do in the studies discussed earlier. Among plant DNA viruses, geminiviruses infect several economically important crops such as cotton,[71,72] tomato,[73] potato,[74] and several legumes,[75,76] as well as ornamentals[77,78] and weeds.[79–81] In the past two decades, molecular biologists and breeders have worked to develop geminivirus-resistant crops.[82,83] Recently, several studies have demonstrated that engineering plant species, including tobacco and *Arabidopsis*, with CRISPR/Cas9 machinery targeting geminiviruses can confer molecular immunity.[14,16,17]

We recently showed that engineering plants to express the CRISPR/Cas9 machinery targeting the TYLCV genome led to interference and degradation of the viral genome, resulting in a significant reduction in viral symptoms.[14] In Ali *et al.* (2015), we showed that *N. benthamiana* plants expressing the CRISPR/Cas9 machinery exhibited resistance against TYLCV, *Beet curly top virus* (BCTV), and *Merremia mosaic virus* (MeMV).[14] Baltes *et al.* and Ji *et al.* demonstrated virus interference activities in *N. benthamiana* against *Bean yellow dwarf virus* (BeYDV) and *Beet severe curly top virus* (BSCTV), respectively.[16,17] Ji *et al.* correlated Cas9 expression with the levels of virus suppression, indicating the need to use a single background transgenic line with optimum expression of Cas9 and sgRNA for practical applications.[17] Baltes *et al.* showed that one sgRNA targeting the BeYDV genome could confer plant resistance without cleavage activity, suggesting that catalytically inactive Cas9 (dCas9) can be used to mediate virus interference, thereby eliminating concerns about off-target activities in the plant genome.[16]

In a follow-up study, we demonstrated that this technology can be used to target and cleave *Cotton leaf curl Kokhran virus* (CLCuKoV) and that by targeting the conserved nonanucleotide sequence, multiple begomoviruses can be targeted simultaneously (CLCuKoV, TYLCV, TYLCSV, MeMV, BCTV-Worland, and BCTV-Logan), conferring broad-spectrum geminivirus resistance.[15] In these studies, *N. benthamiana* plants expressing the CRISPR/Cas9 system displayed considerably reduced viral titers, which abolished or significantly reduced disease symptoms.

We subsequently investigated the differences in efficiency of virus interference when targeting coding vs noncoding sequences.[15] Since invading viruses and host plants exist in an arms race where pathogens rapidly evolve mechanisms to evade plant defenses, we systematically analyzed the ability of geminiviruses to evade the CRISPR/Cas9 machinery. We determined the ability of multiple geminiviruses to evade this machinery by targeting coding and noncoding sequences, finding that targeting coding sequences led to the generation of viral variants capable of evading the CRISPR/Cas9 machinery. Since these viral variants can have variable replication efficiencies, variants with more proficient replication should predominate over those with lower replication rates under natural field conditions, where mixed infections predominate.

Interestingly, targeting noncoding intergenic sequences led to high levels of virus interference, with no detectable viral escapes from the CRISPR/Cas9 machinery, thereby providing an effective strategy for conferring potentially durable resistance. Given the considerable heterogeneity and recombination ability of geminiviruses and the ability for replication and systemic movement in CRISPR/Cas9-induced variants, strategies could be devised in which the CRISPR/Cas9 system targets the noncoding region to inhibit or prevent viral replication by mutating the essential parts of geminiviruses. Alternatively, targeting multiple regions of the virus genome simultaneously would abrogate the ability of the virus to use the NHEJ repair system, and these cleaved molecules would ultimately be degraded.

6. DEVELOPING RESISTANCE AGAINST RNA VIRUSES

Plant viruses that store and transmit their genetic information in form of RNA are called RNA plant viruses. Several RNA viruses cause devastating yield losses in economically important crops like tomato, potato, cucurbits, etc. Genome-engineering strategies aimed at manipulating host factors in the plant genome have been used to develop resistance against RNA viruses.[84] For example, the CRISPR/Cas9 system is used to modify plant genes that affect plant responses to viral infection, thus generating virus resistance. The translation initiation-like factors *eIF4E* and *eIF(iso)4E* are directly involved in the infection process of RNA viruses.[85] *Arabidopsis* plants with mutations in these translation initiation factors exhibit resistance to *Turnip mosaic virus* (TuMV).[86] Therefore, these targets were mutated using the CRISPR/Cas9 system to develop virus-resistant plants.[84,87] The utility of CRISPR/Cas9 technology for generating novel genetic resistance to the potyvirus TuMV was demonstrated in *Arabidopsis* via deletion of the host factor, *eIF(iso)4E*, which is strictly required for viral survival.[84] Virus resistance was also developed in cucumber (*Cucumis sativus*) by disrupting *eIF4E*, resulting in nontransgenic, heterozygous *eIF4E* mutant plants. These nontransgenic plants exhibited partial resistance to an ipomovirus (*Cucumber vein yellowing virus*) and two potyviruses (*Zucchini yellow mosaic virus* and *Papaya ring spot mosaic virus*-W).[87]

Translation initiation factor genes are prime candidates for host genes that can be targeted,[85] but any host gene encoding a factor that the virus requires is a potential target for modification. This form of recessive resistance could be exploited with the aid of the CRISPR/Cas9 system to create novel resistance alleles in crop plants to protect them against problematic viruses that use host translation initiation factors.

7. OUTLOOK

The CRISPR/Cas9 system has recently emerged as a groundbreaking genome-engineering platform. This system has been used to improve several traits in plants, e.g., engineering biotic stress tolerance (against invading viruses,[14–16,81,83,88] fungus,[89,90] bacteria, and insects), abiotic stress tolerance[91] (against drought,[54] heat, and cold stress), and yield improvement[34] (e.g., by increasing grain weight in maize[54]). Moreover, all of these works can be performed using a "DNA-free," nongenetically modified organisms (GMO) crop production system.[92,93] In addition to crop improvement, the CRISPR/Cas9 system can be used to examine the plant genome, e.g., by creating knockout libraries or by enhancing plant genome imaging[94] (Fig. 4). The CRISPR/Cas9 system has been used for targeted genome editing in diverse plant species, from model plants such as *N. benthamiana* and *Arabidopsis* to economically important plants such as wheat, maize, sorghum, and rice.[94] The specificity of this system is determined by the 20-nucleotide sequence of the sgRNA, allowing for unprecedented, simple genome engineering. Furthermore, the CRISPR/Cas9 system can be used to simultaneously edit multiple genomic targets.[95]

Viruses have long been used for genome engineering, mainly as delivery vehicles for mammalian cells.[6] Plant viruses can potentially be powerful tools for improving the efficiency of genome editing.[12,13] Recent studies have demonstrated the utility and applicability of using viruses for genome editing via CRISPR/Cas9 genome-engineering platforms.[94] In these studies, to generate plants with heritable modifications, it was necessary to generate

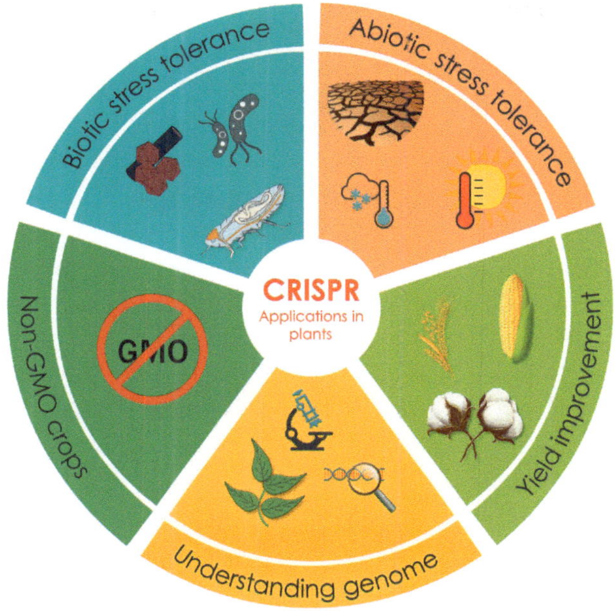

Fig. 4 Applications of CRISPR/Cas9 in plants. CRISPR/Cas9 can be (and has been) used to improve several plant traits, e.g., engineering biotic stress tolerance (resistance against invading viruses, bacteria, and insects), abiotic stress tolerance (resistance against drought, heat, and cold stress), and yield improvement (e.g., by increasing grain weight in wheat, kernel size in corn, and the number of bolls in cotton). Moreover, all of this work can be performed in a "DNA-free," non-GMO crop production system. In addition to crop improvement, the CRISPR/Cas9 system can be used to greatly accelerate fundamental plant research, e.g., by constructing extensive gene knockout libraries or by producing plants with key gene replacements or modifications.

transgenic lines that stably expressed Cas9 and gRNA; progeny with targeted modifications were then recovered in subsequent generations. Since producing transgenic plants is time-consuming, efficient delivery methods are needed to expedite and maximize the utility of this technology for trait discovery and development. The recently developed TRV-mediated CRISPR/Cas9 delivery system[12] can help overcome this hurdle. The use of heterozygous Cas9-overexpressing plants with this simple, versatile genome-editing platform allows plants free of foreign DNA to be generated, which might help overcome the regulatory hurdles that impede the commercialization of genetically engineered plants. TRV has multiple advantages as a delivery vehicle for sgRNA, including its small RNA genome, which is amenable to library construction and does not integrate into the host genome. In addition, TRV can infect germline cells and has a wide host range under both natural and laboratory conditions.[96] However, the limitations in the carrying capacity of TRV and other RNA viruses prevent their use beyond the expression of relatively small SSNs and sgRNAs, and they are unable to efficiently deliver DNA repair templates. Perhaps geminiviruses can be used to overcome the limitations of RNA viruses due to their larger carrying capacity and their ability to produce a DNA replicon capable of acting as a repair template for gene targeting.[97] In an effort to facilitate efficient genome engineering, various viruses have recently been developed to deliver genome-engineering reagents into plants.[12,13,63–66]

Several genome-engineering platforms have been used to engineer plants with molecular immunity against invading viruses. However, the CRISPR/Cas9 system has shown promise for providing immunity against economically important, widespread geminiviruses such as TYLCV. Moreover, targeting the noncoding regions of viruses provides significant interference activity and limits the generation of viral variants that escape the CRISPR/Cas9 machinery and are capable of replication and systemic movement.[15] These developments are exciting but have thus far been limited to laboratory demonstrations. Field trials in multiple locations would be helpful for further evaluating the actual success of this technology under natural conditions. Moreover, previous studies have mainly focused on targeting DNA viruses due to the robust activity of the CRISPR/Cas9 system for cleaving DNA substrates. Novel variants of the CRISPR/Cas system were recently shown to be capable of targeting RNA genomes.[98] These systems are based on the concept that the Cas9 binds with high affinity to single-stranded RNA (ssRNA) targets matching the Cas9-associated gRNA sequence when the PAM is presented in trans as a separate DNA oligonucleotide. Furthermore, PAM-presenting oligonucleotides (PAMmers) stimulate site-specific endonucleolytic cleavage of ssRNA targets; using specially designed PAMmers, Cas9 can be specifically directed to bind or cut RNA targets.[98] Such systems could be applied to target the RNA genomes of single and multiple viruses and to confer resistance against RNA viruses.

One of the salient features of the CRISPR/Cas9 system is that it can be used in a "DNA-free" crop production system to develop nontransgenic plants with desired traits.[99] This is even more practical when endogenous host genes are targeted, since the progeny plants in the following segregating generations can be selected for the desired phenotypes caused by the edited traits, but lacking the Cas9 and sgRNA transgenes.[100] This system has direct application in crop production and variety development.[89,93] An alternate approach to avoid the incorporation of foreign genes into the host genome can be the introduction of CRISPR/Cas9 reagents directly into the cell, for example, by using a Cas9/sgRNA ribonucleoprotein complex.[99,101–103] In this way the final product would be a plant, no different from the one produced as a result of classical plant breeding approaches.[104] The DNA-free mushroom and maize varieties that have been developed using similar CRISPR/Cas9-mediated genome editing approach have been deemed by the US Department of Agriculture to not be subject to their regulatory system.[92,93] This decision may pave the way for these and other plants free of foreign DNAs to enter the marketplace. Overall these strategies can be used to devise a variety development system that might bypass GMO regulations (Fig. 4).[105] However, the actual regulatory status of most genome-edited crops is still under debate.[105]

REFERENCES

1. Stella S, Montoya G. The genome editing revolution: a CRISPR-Cas TALE off-target story. *Bioessays*. 2016;38(S1):S4–S13.
2. Weeks DP, Spalding MH, Yang B. Use of designer nucleases for targeted gene and genome editing in plants. *Plant Biotechnol J*. 2016;14(2):483–495.
3. Price AA, Grakoui A, Weiss DS. Harnessing the prokaryotic adaptive immune system as a eukaryotic antiviral defense. *Trends Microbiol*. 2016;24(4):294–306.
4. Mahfouz MM, Piatek A, Stewart Jr CN. Genome engineering via TALENs and CRISPR/Cas9 systems: challenges and perspectives. *Plant Biotechnol J*. 2014;12(8): 1006–1014.
5. Weeks DP, Jiang W. Genome editing in plants using CRISPR/Cas9/sgRNA technologies. Paper presented at In Vitro Cellular & Developmental Biology-Animal, 2014.
6. Chen X, Goncalves MAFV. Engineered viruses as genome editing devices. *Mol Ther*. 2016;24(3):447–457.
7. Ellis J, Bernstein A. Gene targeting with retroviral vectors: recombination by gene conversion into regions of nonhomology. *Mol Cell Biol*. 1989;9(4):1621–1627.
8. Lombardo A, Genovese P, Beausejour CM, et al. Gene editing in human stem cells using zinc finger nucleases and integrase-defective lentiviral vector delivery. *Nat Biotechnol*. 2007;25(11):1298–1306.
9. Wang Q, Taylor MW. Correction of a deletion mutant by gene targeting with an adenovirus vector. *Mol Cell Biol*. 1993;13(2):918–927.
10. Russell DW, Hirata RK. Human gene targeting by viral vectors. *Nat Genet*. 1998;18(4):325–330.
11. Hendrie PC, Russell DW. Gene targeting with viral vectors. *Mol Ther*. 2005;12(1): 9–17.
12. Ali Z, Abul-faraj A, Li L, et al. Efficient virus-mediated genome editing in plants using the CRISPR/Cas9 system. *Mol Plant*. 2015;8(8):1288–1291.
13. Cermak T, Baltes NJ, Cegan R, Zhang Y, Voytas DF. High-frequency, precise modification of the tomato genome. *Genome Biol*. 2015;16:232.
14. Ali Z, Abulfaraj A, Idris A, Ali S, Tashkandi M, Mahfouz MM. CRISPR/Cas9-mediated viral interference in plants. *Genome Biol*. 2015;16(1):238.
15. Ali Z, Ali S, Tashkandi M, Zaidi SS, Mahfouz MM. CRISPR/Cas9-mediated immunity to geminiviruses: differential interference and evasion. *Sci Rep*. 2016;6:26912.

16. Baltes NJ, Hummel AW, Konecna E, et al. Conferring resistance to geminiviruses with the CRISPR–Cas prokaryotic immune system. *Nat Plants*. 2015;1(10):15145.
17. Ji X, Zhang H, Zhang Y, Wang Y, Gao C. Establishing a CRISPR-Cas-like immune system conferring DNA virus resistance in plants. *Nat Plants*. 2015;1(10):15144.
18. Li T, Liu B, Spalding MH, Weeks DP, Yang B. High-efficiency TALEN-based gene editing produces disease-resistant rice. *Nat Biotechnol*. 2012;30(5):390–392.
19. Aouida M, Li L, Mahjoub A, et al. Transcription activator-like effector nucleases mediated metabolic engineering for enhanced fatty acids production in *Saccharomyces cerevisiae*. *J Biomed Bioeng*. 2015;120(4):364–371.
20. Li T, Huang S, Jiang WZ, et al. TAL nucleases (TALNs): hybrid proteins composed of TAL effectors and FokI DNA-cleavage domain. *Nucleic Acids Res*. 2011;39(1): 359–372.
21. Li T, Huang S, Zhao X, et al. Modularly assembled designer TAL effector nucleases for targeted gene knockout and gene replacement in eukaryotes. *Nucleic Acids Res*. 2011;39:6315–6325. http://dx.doi.org/10.1093/nar/gkr188.
22. Li T, Liu B, Chen CY, Yang B. TALEN-mediated homologous recombination produces site-directed DNA base change and herbicide-resistant rice. *J Genet Genomics*. 2016;43:297–305.
23. Li T, Liu B, Chen CY, Yang B. TALEN utilization in rice genome modifications. *Methods*. 2014;69(1):9–16.
24. Brazelton VA, Zarecor S, Wright DA, et al. A quick guide to CRISPR sgRNA design tools. *GM Crops Food*. 2015;6(4):266–276.
25. Eid A, Mahfouz MM. Genome editing: the road of CRISPR/Cas9 from bench to clinic. *Exp Mol Med*. 2016;48:e265.
26. Wright AV, Nunez JK, Doudna JA. Biology and applications of CRISPR systems: harnessing nature's toolbox for genome engineering. *Cell*. 2016;164(1–2):29–44.
27. Voytas DF. Plant genome engineering with sequence-specific nucleases. *Annu Rev Plant Biol*. 2013;64(1):327–350.
28. Shukla VK, Doyon Y, Miller JC, et al. Precise genome modification in the crop species *Zea mays* using zinc-finger nucleases. *Nature*. 2009;459(7245):437–441.
29. Townsend JA, Wright DA, Winfrey RJ, et al. High-frequency modification of plant genes using engineered zinc-finger nucleases. *Nature*. 2009;459(7245):442–445.
30. Li L, Atef A, Piatek A, et al. Characterization and DNA-binding specificities of Ralstonia TAL-like effectors. *Mol Plant*. 2013;6(4):1318–1330.
31. Ji Z, Ji C, Liu B, Zou L, Chen G, Yang B. Interfering TAL effectors of Xanthomonas oryzae neutralize R-gene-mediated plant disease resistance. *Nat Commun*. 2016;7:13435.
32. Piatek A, Mahfouz MM. Targeted genome regulation via synthetic programmable transcriptional regulators. *Crit Rev Biotechnol*. 2016;19:1–12.
33. Aouida M, Piatek MJ, Bangarusamy DK, Mahfouz MM. Activities and specificities of homodimeric TALENs in *Saccharomyces cerevisiae*. *Curr Genet*. 2014;60(2):61–74.
34. Jiang WZ, Henry IM, Lynagh PG, Comai L, Cahoon EB, Weeks DP. Significant enhancement of fatty acid composition in seeds of the allohexaploid, *Camelina sativa*, using CRISPR/Cas9 gene editing. *Plant Biotechnol J*. 2017, (in press). http://dx.doi.org/10.1111/pbi.12663.
35. Jiang W, Brueggeman AJ, Horken KM, Plucinak TM, Weeks DP. Successful transient expression of Cas9 and single guide RNA genes in *Chlamydomonas reinhardtii*. *Eukaryot Cell*. 2014;13(11):1465–1469.
36. Harms DW, Quadros RM, Seruggia D, et al. Mouse genome editing using the CRISPR/Cas system. *Curr Protoc Hum Genet*. 2014;83(15.7):15.7.1–15.7.27. http://dx.doi.org/10.1002/0471142905.hg1507s83.

37. Housden BE, Perrimon N. Comparing CRISPR and RNAi-based screening technologies. *Nat Biotechnol.* 2016;34(6):621–623.
38. Barrangou R, Doudna JA. Applications of CRISPR technologies in research and beyond. *Nat Biotechnol.* 2016;34(9):933–941.
39. Horvath P, Barrangou R. CRISPR/Cas, the immune system of bacteria and archaea. *Science.* 2010;327(5962):167–170.
40. Barrangou R, Fremaux C, Deveau H, et al. CRISPR provides acquired resistance against viruses in prokaryotes. *Science.* 2007;315(5819):1709–1712.
41. Gasiunas G, Barrangou R, Horvath P, Siksnys V. Cas9–crRNA ribonucleoprotein complex mediates specific DNA cleavage for adaptive immunity in bacteria. *Proc Natl Acad Sci.* 2012;109(39):E2579–E2586.
42. Horvath P, Romero DA, Coûté-Monvoisin A-C, et al. Diversity, activity, and evolution of CRISPR loci in Streptococcus thermophilus. *J Bacteriol.* 2008;190(4): 1401–1412.
43. Mojica F, Diez-Villasenor C, Garcia-Martinez J, Almendros C. Short motif sequences determine the targets of the prokaryotic CRISPR defence system. *Microbiology.* 2009;155(3):733–740.
44. Datsenko KA, Pougach K, Tikhonov A, Wanner BL, Severinov K, Semenova E. Molecular memory of prior infections activates the CRISPR/Cas adaptive bacterial immunity system. *Nat Commun.* 2012;3:945.
45. Sapranauskas R, Gasiunas G, Fremaux C, Barrangou R, Horvath P, Siksnys V. The Streptococcus thermophilus CRISPR/Cas system provides immunity in *Escherichia coli*. *Nucleic Acids Res.* 2011;39:9275–9282. http://dx.doi.org/10.1093/nar/gkr606.
46. Jiang W, Zhou H, Bi H, Fromm M, Yang B, Weeks DP. Demonstration of CRISPR/Cas9/sgRNA-mediated targeted gene modification in Arabidopsis, tobacco, sorghum and rice. *Nucleic Acids Res.* 2013;41(20):e188. http://dx.doi.org/10.1093/nar/gkt780.
47. Ran FA, Hsu PD, Wright J, Agarwala V, Scott DA, Zhang F. Genome engineering using the CRISPR-Cas9 system. *Nat Protocols.* 2013;8(11):2281–2308.
48. Zhou H, Liu B, Weeks DP, Spalding MH, Yang B. Large chromosomal deletions and heritable small genetic changes induced by CRISPR/Cas9 in rice. *Nucleic Acids Res.* 2014;42(17):10903–10914.
49. Paul 3rd JW, Qi Y. CRISPR/Cas9 for plant genome editing: accomplishments, problems and prospects. *Plant Cell Rep.* 2016;25:1–11.
50. Eid A, Ali Z, Mahfouz MM. High efficiency of targeted mutagenesis in arabidopsis via meiotic promoter-driven expression of Cas9 endonuclease. *Plant Cell Rep.* 2016;35: 1555–1558.
51. Jiang W, Yang B, Weeks DP. Efficient CRISPR/Cas9-mediated gene editing in *Arabidopsis thaliana* and inheritance of modified genes in the T2 and T3 generations. *PLoS One.* 2014;9(6):e99225.
52. Zhang Y, Liang Z, Zong Y, et al. Efficient and transgene-free genome editing in wheat through transient expression of CRISPR/Cas9 DNA or RNA. *Nat Commun.* 2016;7:12617.
53. Char SN, Neelakandan AK, Nahampun H, et al. An Agrobacterium-delivered CRISPR/Cas9 system for high-frequency targeted mutagenesis in maize. *Plant Biotechnol J.* 2017;15:257–268.
54. Shi J, Gao H, Wang H, et al. ARGOS8 variants generated by CRISPR-Cas9 improve maize grain yield under field drought stress conditions. *Plant Biotechnol J.* 2017;15:207–216.
55. Mahfouz MM, Cardi T, Neal SC. Next-generation precision genome engineering and plant biotechnology. *Plant Cell Rep.* 2016;35(7):1397–1399.
56. Piatek A, Ali Z, Baazim H, et al. RNA-guided transcriptional regulation in planta via synthetic dCas9-based transcription factors. *Plant Biotechnol J.* 2015;13(4):578–589.

57. Aouida M, Eid A, Ali Z, et al. Efficient fdCas9 synthetic endonuclease with improved specificity for precise genome engineering. *PLoS One*. 2015;10(7):e0133373.
58. Xu S, Cao S, Zou B, et al. An alternative novel tool for DNA editing without target sequence limitation: the structure-guided nuclease. *Genome Biol*. 2016;17(1):186.
59. Marton I, Zuker A, Shklarman E, et al. Nontransgenic genome modification in plant cells. *Plant Physiol*. 2010;154(3):1079–1087.
60. Ali Z, Abul-Faraj A, Piatek M, Mahfouz MM. Activity and specificity of TRV-mediated gene editing in plants. *Plant Signal Behav*. 2015;10(10):e1044191.
61. Shivprasad S, Pogue GP, Lewandowski DJ, et al. Heterologous sequences greatly affect foreign gene expression in tobacco mosaic virus-based vectors. *Virology*. 1999;255(2): 312–323.
62. Choi IR, Stenger DC, Morris TJ, French R. A plant virus vector for systemic expression of foreign genes in cereals. *Plant J*. 2000;23(4):547–555.
63. Gil-Humanes J, Wang Y, Liang Z, et al. High efficiency gene targeting in hexaploid wheat using DNA replicons and CRISPR/Cas9. *Plant J*. 2016. http://dx.doi.org/10.1111/tpj.13446.
64. Butler NM, Baltes NJ, Voytas DF, Douches DS. Geminivirus-mediated genome editing in potato (*Solanum tuberosum* L.) using sequence-specific nucleases. *Front Plant Sci*. 2016;7:1045.
65. Butler NM, Atkins PA, Voytas DF, Douches DS. Generation and inheritance of targeted mutations in potato (*Solanum tuberosum* L.) using the CRISPR/Cas system. *PLoS One*. 2015;10(12):e0144591.
66. Baltes NJ, Gil-Humanes J, Cermak T, Atkins PA, Voytas DF. DNA replicons for plant genome engineering. *Plant Cell*. 2014;26(1):151–163.
67. Chen W, Qian Y, Wu X, Sun Y, Wu X, Cheng X. Inhibiting replication of begomoviruses using artificial zinc finger nucleases that target viral-conserved nucleotide motif. *Virus Genes*. 2014;48(3):494–501.
68. Mori T, Takenaka K, Domoto F, Aoyama Y, Sera T. Inhibition of binding of tomato yellow leaf curl virus rep to its replication origin by artificial zinc-finger protein. *Mol Biotechnol*. 2013;54(2):198–203.
69. Sera T. Inhibition of virus DNA replication by artificial zinc finger proteins. *J Virol*. 2005;79:2614–2619.
70. Cheng X, Li F, Cai J, et al. Artificial TALE as a convenient protein platform for engineering broad-spectrum resistance to begomoviruses. *Virus*. 2015;7(8): 4772–4782.
71. Zaidi SSA, Iqbal Z, Amin I, Mansoor S. First report of tomato leaf curl Gujarat virus, a bipartite begomovirus on cotton showing leaf curl symptoms in Pakistan. *Plant Dis*. 2015;99(11):1655–1656.
72. Zaidi SS, Shafiq M, Amin I, et al. Frequent occurrence of tomato leaf curl New Delhi virus in cotton leaf curl disease affected cotton in Pakistan. *PLoS One*. 2016;11(5), e0155520.
73. Zaidi SS, Martin DP, Amin I, Farooq M, Mansoor S. Tomato leaf curl New Delhi virus: a widespread bipartite begomovirus in the territory of monopartite begomoviruses. *Mol Plant Pathol*. 2017, (in press). http://dx.doi.org/10.1111/mpp.12481.
74. Hameed A, Tahir MN, Amin I, Mansoor S. First report of Tomato leaf curl New Delhi virus and a Tomato yellow leaf curl Thailand betasatellite causing severe leaf curl disease of potato in Pakistan. *Plant Dis*. 2017, (in press). http://dx.doi.org/10.1094/PDIS-09-16-1335-PDN.
75. Zaidi SS, Amin I, Iqbal Z, Akhtar KP, Scheffler BE, Mansoor S. *Sesbania bispinosa*, a new host of a begomovirus-betasatellite complex in Pakistan. *Can J Plant Pathol*. 2016;38(1):107–111.

76. Zaidi SS, Shakir S, Farooq M, Amin I, Mansoor S. First report of a novel strain of Tomato yellow leaf curl virus causing yellow leaf curl disease on cluster bean in Pakistan. *Plant Dis.* 2017, (in press). http://dx.doi.org/10.1094/PDIS-12-16-1859-PDN.
77. Khatri S, Nahid N, Fauquet CM, Mubin M, Nawaz-ul-Rehman MS. A betasatellite-dependent begomovirus infects ornamental rose: characterization of begomovirus infecting rose in Pakistan. *Virus Genes.* 2014;49(1):124–131.
78. Mahesh YS, Shankarappa KS, Rangaswamy KT, Prameela HA, Aswathanarayana DS, Maruthi MN. Detection of a begomovirus associated with leaf curl disease of ornamental croton (Codieum variegatum) in southern India. *J Plant Pathol.* 2008;90:143.
79. Zaidi SS, Shakir S, Farooq M, Amin I, Mansoor S. First report of Alternanthera yellow vein virus from *Eclipta prostrata* in Pakistan. *Plant Dis.* 2017;101(1):266.
80. Zaidi SS, Sheakir S, Maliek HJ, Farooq M, Amin I, Mansoor S. First report of Tomato leaf curl New Delhi virus on Calotropis procera in Pakistan. *Plant Dis.* 2017, (in press). http://dx.doi.org/10.1094/PDIS-10-16-1539-PDN.
81. Zaidi SS, Mansoor S, Ali Z, Tashkandi M, Mahfouz MM. Engineering plants for Geminivirus resistance with CRISPR/Cas9 system. *Trends Plant Sci.* 2016;21(4): 279–281.
82. Zaidi SS, Briddon RW, Mansoor S. Engineering dual Begomovirus-Bemisia tabaci resistance in plants. *Trends Plant Sci.* 2017;22(1):6–8.
83. Zaidi SS, Tashkandi M, Mansoor S, Mahfouz MM. Engineering plant immunity: using CRISPR/Cas9 to generate virus resistance. *Front Plant Sci.* 2016;7:1673.
84. Pyott DE, Sheehan E, Molnar A. Engineering of CRISPR/Cas9-mediated potyvirus resistance in transgene-free Arabidopsis plants. *Mol Plant Pathol.* 2016;17(8):1276–1288.
85. Sanfacon H. Plant translation factors and virus resistance. *Virus.* 2015;7(7):3392–3419.
86. Lellis AD, Kasschau KD, Whitham SA, Carrington JC. Loss-of-susceptibility mutants of *Arabidopsis thaliana* reveal an essential role for eIF(iso)4E during potyvirus infection. *Curr Biol.* 2002;12(12):1046–1051.
87. Chandrasekaran J, Brumin M, Wolf D, et al. Development of broad virus resistance in non-transgenic cucumber using CRISPR/Cas9 technology. *Mol Plant Pathol.* 2016; 17(7):1140–1153.
88. Subburaj S, Tu L, Jin Y-T, Bae S, Lee G-J. Development of Potyvirus resistance in non-transgenic tomato plants using CRISPR/Cas9 technology. Paper Presented at Korea Breeding Society Symposium 2016 Korean Breeding Society—Next Generation BG21 Project—GSP Project Joint Symposium. 2016. http://db.koreascholar.com/article.aspx?code=317459.
89. Wang Y, Cheng X, Shan Q, et al. Simultaneous editing of three homoeoalleles in hexaploid bread wheat confers heritable resistance to powdery mildew. *Nat Biotechnol.* 2014;32(9):947–951.
90. Wang F, Wang C, Liu P, et al. Enhanced rice blast resistance by CRISPR/Cas9-targeted mutagenesis of the ERF transcription factor Gene OsERF922. *PLoS One.* 2016;11(4):e0154027.
91. Osakabe Y, Watanabe T, Sugano SS, et al. Optimization of CRISPR/Cas9 genome editing to modify abiotic stress responses in plants. *Sci Rep.* 2016;6:26685.
92. Waltz E. CRISPR-edited crops free to enter market, skip regulation. *Nat Biotechnol.* 2016;34(6):582.
93. Waltz E. Gene-edited CRISPR mushroom escapes US regulation. *Nature.* 2016; 532(7599):293.
94. Noman A, Aqeel M, He S. CRISPR-Cas9: tool for qualitative and quantitative plant genome editing. *Front Plant Sci.* 2016;7(1740):1740.
95. Cong L, Ran FA, Cox D, et al. Multiplex genome engineering using CRISPR/Cas systems. *Science.* 2013;339(6121):819–823.

96. Senthil-Kumar M, Mysore KS. Tobacco rattle virus-based virus-induced gene silencing in Nicotiana benthamiana. *Nat Protocols*. 2014;9(7):1549–1562.
97. Richter KS, Kleinow T, Jeske H. Somatic homologous recombination in plants is promoted by a geminivirus in a tissue-selective manner. *Virology*. 2014;452–453:287–296.
98. O'Connell MR, Oakes BL, Sternberg SH, East-Seletsky A, Kaplan M. Doudna JA. Programmable RNA recognition and cleavage by CRISPR/Cas9. *Nature*. 2014; 516(7530):263–266.
99. Woo JW, Kim J, Kwon SI, et al. DNA-free genome editing in plants with preassembled CRISPR-Cas9 ribonucleoproteins. *Nat Biotechnol*. 2015;33(11):1162–1164.
100. Kanchiswamy CN. DNA-free genome editing methods for targeted crop improvement. *Plant Cell Rep*. 2016;35(7):1469–1474.
101. Kim J, Kim J-S. Bypassing GMO regulations with CRISPR gene editing. *Nat Biotechnol*. 2016;34(10):1014–1015.
102. Malnoy M, Viola R, Jung M-H, et al. DNA-free genetically edited grapevine and apple protoplast using CRISPR/Cas9 ribonucleoproteins. *Front Plant Sci*. 2016;7:1904.
103. Liang Z, Chen K, Li T, et al. Efficient DNA-free genome editing of bread wheat using CRISPR/Cas9 ribonucleoprotein complexes. *Nat Commun*. 2017;8:14261.
104. Voytas DF, Gao C. Precision genome engineering and agriculture: opportunities and regulatory challenges. *PLoS Biol*. 2014;12(6):e1001877.
105. Wolt JD, Wang K, Yang B. The regulatory status of genome-edited crops. *Plant Biotechnol J*. 2016;14(2):510–518.

CHAPTER ELEVEN

Use of CRISPR/Cas9 for Symbiotic Nitrogen Fixation Research in Legumes

Longlong Wang, Longxiang Wang, Yu Zhou, Deqiang Duanmu[1]
State Key Laboratory of Agricultural Microbiology, College of Life Science and Technology, Huazhong Agricultural University, Wuhan, China
[1]Corresponding author: e-mail address: duanmu@mail.hzau.edu.cn

Contents

Abstract

Nitrogen-fixing rhizobia have established a symbiotic relationship with the legume family through more than 60 million years of evolution. Hundreds of legume host genes are involved in the SNF (symbiotic nitrogen fixation) process, such as recognition of the bacterial partners, nodulation signaling and nodule development, maintenance of highly efficient nitrogen fixation within nodules, regulation of nodule numbers, and nodule senescence. However, investigations of SNF-related gene functions and dissecting molecular mechanisms of the complicated signaling crosstalk on a genomic scale were significantly restricted by insufficient mutant resources of several

Progress in Molecular Biology and Translational Science, Volume 149
ISSN 1877-1173
http://dx.doi.org/10.1016/bs.pmbts.2017.03.010

representative model legumes. Targeted genome-editing technologies, including ZFNs, TALENs, and CRISPR–Cas systems, have been developed in recent years and rapidly revolutionized biological research in many fields. These technologies were also applied to legume plants, and significant progress has been made in the last several years. Here, we summarize the applications of these genome-editing technologies, especially CRISPR–Cas9, toward the study of SNF in legumes, which should greatly advance our understanding of the basic mechanisms underpinning the legume–rhizobia interactions and guide the engineering of the SNF pathway into nonlegume crops to reduce the dependence on the use of nitrogen fertilizers for sustainable development of modern agriculture.

1. INTRODUCTION

Between the 1930s and the late 1960s, development of high-yielding varieties of cereals, in association with chemical fertilizers and new methods of cultivation, irrigation, and modernization of management techniques, resulted in a steady increase in the productivity of major crops such as rice, maize, and wheat.[1] However, the excessive use of synthetic fertilizers, most notably the nitrogen and phosphorus fertilizers, has consumed an enormous amount of nonrenewable resources during production processes and unfortunately has resulted in pollution of soils, groundwaters, and lake systems worldwide, especially in developing countries such as China and India. With an estimated world population reaching more than 9 billion by 2050, the annual cereal production must increase by at least another 70% compared to the level in 2005 in the next three decades to meet the global demand for food, feed, and fiber.[2] Undoubtedly, one of the ensuing key questions is: how can we control or even reduce the usage of synthetic nitrogen fertilizers to achieve the sustainable development of the Next Green Revolution?

Atmospheric nitrogen gas (N_2) must be processed into organic nitrogen to be taken up by plants. Compared with industrial nitrogen fixation by the energy-consuming Haber–Bosch process, which now produces ~120 million tons of nitrogen fertilizer per year, biological nitrogen fixation (BNF) is an environment-friendly alternative that contributes more than 50% to the global nitrogen fixation (estimated around 413 million tons in 2010).[3] The BNF depends on the activity of nitrogenase, a metalloprotein complex found in a wide variety of free-living or symbiotic bacteria known as diazotrophs. The symbiotic bacteria such as rhizobium can establish a mutualistic relationship with many legume species, producing ammonia in exchange for carbohydrates within the nodule, a specialized root-derived organ that has evolved to host highly efficient nitrogen fixation.[4] The legumes are thus ideal agricultural crops as their demand for nitrogen fertilizer is dramatically reduced. Indeed, soybeans and other legumes are routinely used in rotations with cereal crops to nourish poor soils that have been depleted of nitrogen.

The legume family (Fabaceae or Leguminosae) is the third largest and economically important family of flowering plants, consisting of ~19,000 known species.[5] Several legumes have a long history of use as crop plants, forage, and green manure, i.e., *Glycine max* (soybean), *Arachis hypogaea* (peanut), *Phaseolus vulgaris* (beans), *Medicago sativa* (alfalfa), and *Astragalus sinicus* (Chinese milk vetch). In the last two decades, both applied and basic scientific investigations have been intensively performed on several model legumes, such as *Medicago truncatula*, *Lotus japonicus*, and *Glycine max*, whose genomes have been recently sequenced completely.[6–8] In this chapter, we summarize recent knowledge regarding symbiotic nitrogen fixation (SNF) and discuss the applications of clustered regularly interspaced short palindromic repeats (CRISPR)/Cas9 genome-editing technology in legume research.

2. SYMBIOTIC NITROGEN FIXATION

2.1 Interactions Between Legumes and Rhizobia

Under low-nitrogen conditions, specific flavonoids are produced and secreted from legume roots into the rhizosphere.[9] Rhizobia perceive these chemical signals and activate expression of a suite of bacterial nodulation genes, which leads to the synthesis of nodulation factors (NFs), a mixture of lipochitooligosaccharide (LCO) molecules that consist of a chitin backbone with various functional group substitutions.[10] NFs activate nodulation signaling in legumes after being recognized by a specific class of receptor kinases (Table 1). Simultaneously, the rhizobia can infect legume roots through entry into surface cracks or via infection threads. This infection leads to stimulation of cell division in the root cortex and meristem formation.[9,10,53]

Flavonoids and NFs are believed to play major roles in determining the host specificity during legume–rhizobia interaction. However, an NF-independent pathway also exists to mediate the recognition of rhizobia by legumes. Okazaki *et al.*[53a] found that an NF-deficient mutant of *Bradyrhizobium elkanii* could induce nitrogen-fixing nodules in soybean

Table 1 Symbiotic Nitrogen Fixation-Associated Genes in Representative Legumes

	Homologous Genes				
	Lj	**Mt**	**Gm**	**Gene Function/Description**	**References**
Signaling receptors	*LjNFR1*	*MtLYK3*	*GmNFR1α*, *GmNFR1β*	An LysM receptor kinase, NF receptor	11–14
	LjNFR5	*MtNFP*	*GmNFR5α*, *GmNFR5β*	An LysM receptor kinase, NF receptor	11,15–17
	LjEPR3	—	—	An LysM receptor kinase, EPS receptor	18
Nod signaling and nodule organogenesis	*LjSYMRK*	*MtDMI2*	*GmSYMRKα*, *GmSYMRKβ*	An LRR receptor kinase required for both root nodule symbiosis and mycorrhizal symbiosis	19–21
	LjPOLLUX/ *LjCASTOR*	*MtDMI1*	—	Ca^{2+}-gated ion channel required for the generation of Ca^{2+} spiking	22,23
	—	*MtCNGC15s*	—	Nuclear-localized cyclic nucleotide-gated channels required for nuclear Ca^{2+} spiking	24
	—	*MtMCA8*	—	A SERCA-type calcium ATPase required for nuclear Ca^{2+} spiking	25
	LjCCaMK	*MtDMI3*	—	Ca^{2+}/calmodulin-dependent kinase, putative Ca^{2+} signal decoder	26,27
	LjCYCLOPS	*MtIPD3*	—	Transcription factor required for infection threads elongation	28,29
	LjNIN	*MtNIN*	*GmNIN-1a*, *GmNIN-1b*, *GmNIN-2a*, *GmNIN-2b*	Core transcription factor with multiple functions in nodule organogenesis	30–32

	LjNSP1	*MtNSP1*	*GmNSP1a*, *GmNSP1b*	GRAS domain containing transcription factor activating downstream gene expression	32–34
	LjNSP2	*MtNSP2*	*GmNSP2a*, *GmNSP2b*	GRAS domain containing transcription factor activating downstream gene expression	32,33,35
	LjERN1	*MtERN1*	*GmERN1a*, *GmERN1b*	AP2-ERF domain containing transcription factor required for nodule development	32,36,37
	LjLHK1	*MtCRE1*	—	Cytokinin receptor kinase required for nodule organogenesis	38–40
	LjNF-YA1	*HAP2-1 NF-YA1*	*GmNF-YA1a/b*	CCAAT box-binding transcription factor required for nodule meristem development	41–43
Autoregulation of nodulation (AON)	*LjHAR1*	*MtSUNN*	*GmNARK*	An LRR receptor kinase required for systematic regulation of nodule numbers	44–47
	LjCLE-RS1/2	*MtCLE12/13*	*GmRIC1/2*	Root-derived small peptides for long-distance signaling	48–51
	LjTML	—	—	An F-box protein negatively regulates nodule numbers	52

"—" indicates no publications on corresponding homologous genes. For simplicity, only well-characterized genes involved in NF perception, Nod signaling, nodule organogenesis, and AON were listed in this table. *Gm*, *Glycine max*; *Lj*, *Lotus japonicus*; *Mt*, *Medicago truncatula*.

through the type III secretion system, indicating that some unknown effectors secreted by *B. elkanii* are injected into plant cells and activate host symbiosis signaling to induce nodule formation.

2.2 Signaling Mechanisms and Nodule Organogenesis

Legumes rely on plasma membrane-localized receptor kinases to respond to rhizobia Nod factors. In *L. japonicus*, two LysM domain-containing receptor kinases (NFR1, NFR5) can form a heterodimer complex and bind NF directly with a K_d value in the nanomolar range.[11,15,54] Recently, a putative exopolysaccharide (EPS) receptor, EPR3, was also identified in *Lotus*. EPR3 might be involved in downregulation of plant defense reaction and in promoting rhizobial infection through binding bacterial EPSs.[18] Once the Nod factor is perceived, signals will be transmitted to the downstream leucine-rich repeat receptor-like kinase (SYMRK) that is essential for both root nodule symbiosis and mycorrhizal symbiosis.[19] Homologous genes encoding these receptor kinases were also characterized in other legumes, such as *M. truncatula* (*MtLYK3*, *MtNFP*, and *DMI2*) and *G. max* (*GmNFR1α/β*, *GmNFR5α/β*, and *GmSymRKα/β*) (Table 1).

Activation of cell membrane receptors rapidly triggers oscillations of the spiking of nuclear and perinuclear calcium (Ca^{2+}) concentrations, partially through a mevalonate-dependent pathway.[55] Ca^{2+} is a key secondary messenger during nodulation signaling, and periodic Ca^{2+} spiking is maintained via the changes in the membrane potential. Several genes have been found to be essential for sustaining the Ca^{2+} spiking, including the potassium-permeable channel proteins (POLLUX/CASTOR), Ca^{2+}-dependent adenosine triphosphatase (MCA8) and Ca^{2+}-permeable channels (cyclic nucleotide-gated channels, CNGC15).[24,25] CCaMK, a calcium calmodulin-dependent protein kinase, decodes the Ca^{2+} signal and phosphorylates its interaction partner CYCLOPS.[56,57] CYCLOPS is a DNA-binding transcriptional activator, which directly binds to the NIN (nodule inception) promoter and activates its expression.[58] GRAS-type transcription factors, nodulation signaling pathway 1 (NSP1) and NSP2, can form a complex and stimulate the expression of NIN.[33,59] As a core transcription factor, NIN plays multiple functions during symbiosis, i.e., formation of infection threads,[60] homeostatic regulation of nodulation,[61] cytokinin responses,[62] etc.

Several phytohormones are known to play key roles in nodule organogenesis, including cytokinin, ethylene, auxin, gibberellin, salicylic acid, abscisic acid, and jasmonic acid.[63] Among these, cytokinin serves as a positive regulator by orchestrating signaling events that promote nodule primordium formation and rhizobial colonization. In *L. japonicus*, this process is mediated by Lotus histidine kinase 1 (LHK1), the main cytokinin receptor that functions partially redundantly with LHK1a and LHK3.[64] A balance between auxin and cytokinin signaling is critical for development of many secondary organs, i.e., lateral branches and lateral roots. During nodule organogenesis, local auxin accumulation caused by inhibition of polar auxin transport is a prerequisite for nodule primordia initiation.[63] Several recent transcriptomic studies revealed the mutual regulation by auxin and cytokinin of the biosynthesis and metabolism of these two hormones and their downstream signaling factors.[65] Other hormones, such as abscisic acid, jasmonic acid, ethylene and salicylic acid, are involved in inhibitory regulation of nodulation through a complex network consisting of several signal transduction pathways.[63] The intimate crosstalk between all these phytohormones plays an indispensable role in legume–rhizobia interactions.[63]

Due to the energy-intensive nature of nitrogen fixation, nodule numbers are under strict control by the legume hosts via a mechanism called autoregulation of nodulation (AON).[44] The production of CLE-RS1/2 peptides is trigged by rhizobial infection of *L. japonicus*.[61] These root-derived small peptides are transported to the shoots and perceived by the receptor kinase HAR1. The *har1* mutant displays a hypernodulation phenotype, indicating the dysfunction of AON.[44] Downstream of the CLE-RS1/2-HAR1 signaling pathway, shoot-derived cytokinin can also systemically regulate root nodulation.[48] In the soybean root, miR172c was found to be negatively regulated by NARK (the homologous protein of HAR1 in *G. max*), probably through the inhibition of shoot-derived cytokinin signaling.[66] Another hypernodulation mutant, too much love (*tml*), was also characterized in *L. japonicus*. The TML protein contains an F-box domain and is proposed to function downstream of HAR1 as a receptor or mediator of unknown mobile systematic signals.[52] Besides these genetic components involved in regulation of AON, several environmental parameters also contribute to AON regulation, such as light, soil nutrients (especially nitrate availability), etc.[67,68]

2.3 Maintenance of Nitrogen Fixation Efficiency

Within the symbiosome, the intracellular organelle consisting of bacteria that continuously differentiate into bacteroids, highly efficient nitrogen fixation is tightly controlled by complicated signal and nutrient exchanges between the microsymbiont and the host. During nutrient exchanges between the two symbiotic partners, several key enzymes and transporters are responsible for the supply of carbon sources and other crucial nutrients. These include the nodule-specific sucrose transporter MtSWEET11[69] and downstream catabolic enzymes, carbonic anhydrase, phosphoenolpyruvate carboxylase, malate dehydrogenase,[70] sulfate transporter LjSST1,[71] citrate transporter LjMATE1,[72] amino acid transporters, etc. In return, nitrogen fixed by the bacteroids is excreted into the plant cytoplasm and rapidly assimilated into amino acids by glutamate synthase (GOGAT) and aspartate aminotransferase. Besides the exchange of fixed carbon and fixed nitrogen, provision of essential elements is also indispensable for the symbiotic microbes, including phosphorus, iron, sulfur, molybdenum, and vanadium.[73] Furthermore, since nitrogenase is extremely sensitive to oxygen, plants and the symbionts have evolved various mechanisms to maintain the microaerobic conditions favorable to nitrogenase, i.e., binding and transport of oxygen by a high concentration of leghemoglobins, rapid consumption of oxygen by bacteria and plant respiration, and specialized cell layers within the nodules that serve as physical barriers against oxygen diffusion, etc.[74]

2.4 Nodule Senescence

On average, nodules on fast-growing herbaceous legumes have a life span of around 10–12 weeks in the absence of stress.[75] Nodule senescence can be triggered by either developmental cues or stress induction. Nodule senescence triggered by stress exhibits distinctive structural and gene expression patterns, underscoring divergent underlying molecular mechanisms for the two distinct types of senescence. In natural conditions, the onset of nodule senescence can be observed as change in nodule color from pink to green due to the pronounced degradation of leghemoglobins.[75] Several other representative markers have also been characterized, including distinct changes in ascorbate and glutathione concentrations,[76] ROS accumulation,[75] and a recently identified cysteine protease and the protease inhibitor, cystatin.[77] Proteolytic activity of a family of conserved cysteine proteases is essential in degrading the symbiotic partners, with cytosolic leghemoglobin and the bacteroid as prime targets.[77,78] Cystatins are cysteine protease inhibitors possibly involved in regulating proteolysis during the programmed cell death that accompanies nodule senescence.[77,79] In addition to the developmental cues, nodule senescence is also induced by dark incubation, treatment with excessive nitrate, and drought stress.[75]

3. CREATING GENE MUTATIONS IN LEGUMES

3.1 Physical and Chemical Mutagenesis

Genome-wide mutant resources are highly valuable for legume functional genomics research. The traditional approach for construction of large-scale mutant libraries mainly relies on physical and chemical mutagenesis techniques. Mutagenized populations induced by fast neutron (FN) radiation were successfully established in soybean. Most mutations resulting from FN bombardment are DNA deletions that range in size from a few base pairs to several megabases[80] (Table 2). Ethyl methanesulfonate (EMS) is also widely used in a variety of legumes, including *G. max*,[81] *L. japonicus*,[88,89] and *M. truncatula*.[87] By extensive screening of ~40,000 EMS-mutagenized *L. japonicus* M2 seeds, 32 mutant lines were isolated with SNF-related phenotypes, including hypernodulating mutants *Ljsym78-1* and *Ljsym78-2*, and nitrogen fixation-deficient mutant *fen1*.[88] Besides these forward genetics

Table 2 Available Mutant Libraries in Model Legumes

Legume Species	Mutagen	Population Size	Website	References
Glycine max	FN	>20,000	https://www.soybase.org/mutants/	80
	EMS	116	—	81
	Tnt1	27	—	82
Lotus japonicus	*LORE1*	121,531	http://lotus.au.dk	83
	EMS	8556	http://revgenuk.jic.ac.uk/order.htm	84
	Tnt1	51	—	85
Medicago truncatula	*Tnt1*	~12,000	http://medicago-mutant.noble.org/	86
	EMS	3162	http://www.inra.fr/legumbase	87

"—" indicates no publicly available information.

methods, TILLING, targeting-induced local lesions in genomes, is a reverse genetic strategy to identify mutations in specific genes from a large number of mutants in mutagenized populations.[87]

Typically, mutants induced by physical or chemical mutagenesis generate genome-wide random DNA alterations. Mutagenesis by these methods is relatively easy to employ with many legumes. The main advantage of these mutagenesis methods is that a large number of mutants can be obtained in a single experiment without employing laborious tissue culture procedures.[88] However, because the gene mutations are untagged when using these mutagens and usually there are multiple mutation sites within the genome, identification, and mapping of genes corresponding to an interesting mutant requires several rounds of backcrossing to clean up the background before time-consuming and costly map-based cloning can be performed—or even before productive whole-genome resequencing can begin.

3.2 Transposon Mutagenesis

In several published studies, insertional mutagenesis was also carried out in a number of legume species based on several artificial transposons or retrotransposons, such as *Tnt1* (the transposable element of *Nicotiana tabacum* cell type 1) and *LORE1* (long terminal repeat element *Lotus retrotransposon 1*). Transposons are engineered into binary vectors and introduced into plant genome using *Agrobacterium tumefaciens*-mediated transformation. The most widely used transposon, *Tnt1*, has been used successfully in many legumes, such as *G. max*,[82] *L. japonicus*,[85] and *M. truncatula*.[86,90] By using *LORE1* retrotransposon, a large mutant population with an excess of 120,000 lines was constructed and deposited into *Lotus* Base, an open-source and free-of-charge integrated information portal to provide mutant information and comprehensive data to legume researchers.[91,92]

Although transposon mutagenesis has been widely used in many legume species, and these genetic resources have significantly accelerated functional genomic research in legumes, the use of only one type of transposon may not be sufficient to construct a saturated gene mutation library due to the nonrandom insertion of the respective transposons.[93] Moreover, larger numbers of inserted copies in each mutant complicate genetic dissection and gene cloning procedures. For example, an average of 4.76 insertions was detected per mutant line among individuals in a *Lotus LORE1* insertion population.[91] More problematically, both *Tnt1* and *LORE1* are derepressed during tissue culture, resulting in a "copy-and-paste" mode of duplication of the DNA elements into other sites in the genome.[90] Thus, it is very difficult to perform functional complementation analysis via stable transformation using these lines.

3.3 Targeted Genome Editing With ZFN and TALEN

Targeted genome editing has revolutionized almost all biological research fields by creating specific mutations that alter or knockout target gene function. With the development of high-throughput and inexpensive genome sequencing technology, an increasing number of legume genomes have been sequenced and the genomic information is now publicly accessible, i.e., *G. max*,[8] *L. japonicus*,[6] *M. truncatula*,[7] and *Aeschynomene evenia*.[94] Zinc finger nucleases (ZFNs), transcription activator-like effector nucleases (TALENs), and CRISPR are among the most sophisticated tools for targeted genome editing.

ZFNs consist of two major parts, the DNA-binding domain composed of a group of four to six zinc finger proteins (ZFP) and the nuclease domain from the *Fok*I restriction enzyme. The ZFP region contains a tandem array of Cys2–His2 fingers, each recognizing approximately 3 bp of DNA. The combined specificity of the individual ZFNs determines the specific 12- to 18-nucleotide DNA sequence to be bound. As *Fok*I needs to form a dimer to cleave DNA double strands, two ZFP-*Fok*1 fusion proteins target to opposite DNA strands on both sides of the cleavage site. Subsequently, the resulting double-strand breaks (DSBs) trigger DNA repair through either of two different mechanisms: homologous recombination (HR) or nonhomologous end joining (NHEJ).[95]

TALENs employ the same principle as ZFNs in that the functional domain for DNA cleavage is contributed by the *Fok*I dimer. However, the DNA-binding specificities of TALENs are based upon designer TALEs based on the DNA recognition code of natural TALEs from plant pathogenic *Xanthomonas*.[96] TALENs consist of a central repeat domain that mediates DNA recognition. Each repeat contains 33–35 amino acids and is capable of binding to one of the four deoxynucleotides, A, G, C, or T. Specificity for binding to a unique nucleotide sequence in a plant genome is dictated by the order in which the nucleotide-specific repeats are arranged in the central domain. Two TALE-nuclease chimeras targeting adjacent sequences within the target gene lead to proper positioning and joining of their *Fok*I domains and creation of a DSB at a precisely predicted location within the genome.[97]

To date, there are only few published studies regarding the applications of ZFNs and TALENs in *G. max*, yet no progress has been reported with other legumes. In soybean, the paralogous genes *DCL4a* and *DCL4b* were successfully knocked out using both hairy root and stable *Agrobacterium*-mediated transformation to deliver ZFNs.[98] Two key genes affecting soybean oil quality, *FAD2-1A* and *FAD2-1B*, were knocked out using TALENs.[99] The complexity of designing DNA-binding modules, the inefficiency of genome targeting, and the prevalent off-target effects of ZFNs and TALENs have significantly restricted their applications in legume research.[100]

4. APPLICATIONS OF CRISPR/Cas9 IN LEGUMES

4.1 CRISPR/Cas9 Principles

CRISPR–Cas system is a recently characterized adaptive immunity system employed by prokaryotes and some archaea to allow resistance to invading viruses and other nucleic acids.[101,102] Based on the phylogenies of *Cas* genes, the organization and sequence of the CRISPR repeats, and the organization of *Cas* operons, the CRISPR system can be divided into three major types (I, II, and III).[103] Among them, type II is the simplest one because only one Cas protein (e.g., Cas9) is required for silencing of foreign DNA. Notably, the type II system is absent from archaea and only present in ~5% bacterial genomes.

To achieve the directed cleavage of the invading foreign DNA, the Cas9 protein is guided to its target by a molecule containing two RNAs, tracrRNA and crRNA, held together by hybridization of a stem-loop structure. A 20 nucleotide sequence in the crRNA ("guide" RNA) is responsible for detecting and binding a specific DNA target as the Cas9/tracrRNA:crRNA complex scans the invading DNA molecule. Proper alignment of the crRNA with its target triggers a conformational rearrangement in the Cas9 molecule and activation of its two nuclease domains to create a DSB. Presence of a DSB, in turn, triggers destruction of the invading molecule. This compact enzymatic machinery and easily manipulatable site-specific DNA targeting make the type II CRISPR system a powerful tool for targeted genome engineering. The prime skill needed is in designing an appropriate sequence-specific guide RNA (gRNA).[104,105] In nature, the "guide" RNA consists of a crRNA spacer sequence at the 5′-end that matches the protospacer DNA sequence of the target locus and a conserved

3′-end tracrRNA scaffold with a special stem-loop structure that is bound by Cas9 nuclease.[106] Importantly, Jinek *et al.* demonstrated that the two RNAs could be combined into a single guide RNA (sgRNA) that was fully active.[104] Once formed, the Cas9/sgRNA complex can make a DSB at its specific target site. As with DSBs created by ZFNs or TALENs, DNA cleavage by Cas9 is followed by either NHEJ- or HR-mediated repair, resulting in gene modifications at the target site.[107] Typically the modifications include small nucleotide insertions or deletions (indels), fragment insertion or deletion, or even DNA fragment (or gene) substitution in the presence of a homologous donor DNA.[106] To date, CRISPR/Cas9-mediated genome editing has been established in many plant species, including *Arabidopsis*,[108] rice,[109] tobacco,[107] tomato,[110] potato[111] and many legumes such as soybean,[112,113] *Lotus*,[114] and *Medicago*[115] (Table 3).

4.2 Legume Genome Editing with CRISPR/Cas9

As for legume plants, CRISPR/Cas9 was first utilized for soybean genome modification using hairy root transformation mediated by *Rhizobium rhizogenes* (formerly *Agrobacterium rhizogenes*). Both endogenous and exogenous genes were effectively disrupted[116,117] (Table 3). The efficiency of genome editing using sgRNAs driven by various endogenous and exogenous RNA polymerase III (pol III) promoters were compared and the results suggested that both the *GmU6-10* promoter[112] and the *GmU6-16g-1* promoter[118] worked more efficiently than *AtU6-26* promoter in transgenic hairy roots. In contrast, the mutation efficiency of TALENs was much lower than that of Cas9/sgRNA system when using the soybean *U6* promoter.[118] More encouragingly, Cas9–sgRNA together with donor DNA fragments was successfully transformed into soybean using the particle bombardment method and homozygous T1 plants with the expected precise DNA knock-in were obtained.[113]

M. truncatula and *L. japonicus* are most commonly used model legumes for SNF studies, largely due to their relatively small genome size and diploid nature compared to the tetraploid soybean. The efficacy of CRISPR/Cas9 system has been validated in these two species using both hairy root transformation[114,119] and stable transformation.[114,115] In *L. japonicus*, up to three homologous SNF genes (leghemoglobins *LjLb1*, *LjLb2*, and *LjLb3*) were efficiently disrupted in the same transgenic line by using two common sgRNAs[114] (Table 3).

Table 3 ZFNs, TALENs, and CRISPR-Mediated Genome Editing in Legumes

Technologies	Legume Species	Target Genes	Transformation Methods	Editing Efficiency (%)	References
ZFNs	*Glycine max*	*DCL4a*, *DCL4b*	18r12-mediated stable transformation	—	98
TALENs	*Glycine max*	*FAD2-1A*, *FAD2-1B*	18r12-mediated stable transformation	21	99
CRISPR	*Glycine max*	*Glyma06g14180*, *Glyma08g02290*, *Glyma12g37050*	K599-mediated hairy root transformation	3.2–9.7 (*AtU-26*) 14.7–20.2 (*GmU6-10*)	112
		GmFEI2, *GmFEI1*, *GmSHR*, *bar*	K599-mediated hairy root transformation	54 (in average)	116
		DD20, *DD43*	Particle bombardment transformation	59–76	113
		01gDDM1, *11gDDM1*, *Glyma04g36150* *miR1509*, *miR1514*, *Glyma06g18790*, *GFP*	K599-mediated hairy root transformation	70 (in average)	117
		GmPDS11, *GmPDS18*	K599-mediated hairy root transformation	11.7–18.1 (*AtU-26*) 43.4–48.1 (*GmU6-16g-1*)	118
	Lotus japonicus	*LjSYMRK*, *LjLbs*	LBA1334-mediated hairy root transformation	59–71	114
			EHA105-mediated stable transformation	35	
	Medicago truncatula	*MtPDS*	AGL1-mediated stable transformation	10.4	115

"—" indicates no statistical data; 18r12, K599, LBA1334, EHA105, and AGL1 represent different *Agrobacterium* species used in plant transformation.

4.3 Guide RNA Design and Expression

The specificity and efficiency of CRISPR-mediated genome modifications are largely determined by the sgRNA component of the Cas9/sgRNA complex. Therefore, it is essential to design appropriate sgRNAs that minimize the potential of creating DSBs at off-target sites in the genome. Two representative online platforms are commonly used for sgRNA design in legumes, CRISPR-PLANT and CRISPR-P. CRISPR-PLANT provides sgRNA spacer sequences with specificity information that are suitable for eight model plants and crops including *G. max* and *M. truncatula.*[120] CRISPR-P is applicable to more than 33 plant genome databases, including *G. max*, *M. truncatula*, and *L. japonicus*. CRISPR-P allows users to search for highly specific target sites and also select restriction enzyme recognition sites for straightforward genotype identification of the putative mutagenized plants. Additionally, CRISPR-P provides off-target prediction for specificity analyses.[121]

Both of the two online resources accommodate the demand for single gene or multigene disruption. For editing of homologous genes simultaneously, one sgRNA is usually not sufficient. The alternative strategy is to design several common sgRNAs targeting the conserved regions in all homologous genes. Use of the common sgRNAs strategy also provides the potential capability to produce single, double, and multiple mutant transgenic plants among the offspring. For example, the three *L. japonicus* leghemoglobin genes (*LjLbs*) share more than 94% identity at DNA level.[74] By using two common sgRNAs, complete disruption of all six alleles was achieved.[114] This approach should be extremely useful for targeting soybean genes since nearly 75% of the genes are present in multiple copies, including many SNF-related genes.[8,16]

Adequate expression levels of multiple sgRNA genes are often equally important for efficient DNA editing. There are two frequently used strategies to simultaneously generate sufficient numbers of different sgRNAs to edit multiple genes or multiple sites within a single gene. One strategy employs a tandem array of sgRNA genes separated from each other by nucleotide sequences that when transcribed into RNA are recognized by tRNA-processing enzymes and converted into single sgRNAs.[122] The second strategy involves tandem assembly of multiple sgRNA genes, each driven independently by one of a number of different U6 or U3 promoters.[123] The tRNA–sgRNA array system does not require cloning of very large DNA fragments, which provides an attractive choice for multiple gene editing. However, this technology is confronted with the limitations that no more than eight tRNA–gRNA repeats can be assembled using the Golden Gate cloning method. Recently, Tang et al. used only one Pol II promoter to control the expression of Cas9 and sgRNAs in a single transcriptional unit (STU), with a hammerhead ribozyme recognition site present between Cas9 and sgRNAs to facilitate processing of multiple sgRNAs.[148] This system should help expand the application of CRISPR-based gene editing to many nonmodel organisms.

4.4 Cas9 Proteins

The level of expression of the Cas9 protein could be a limiting factor affecting the efficiency of gene editing. To obtain high-level Cas9 protein accumulation, it is preferable to optimize the codon usage in a *Cas9* gene to match the codon preference of a particular species. So far, both dicotyledon codon-optimized Cas9[116,118] and human codon-optimized Cas9[114,117] have been successfully used in legume plants with relatively similar efficiency. In several published studies, promoters of endogenous housekeeping genes (i.e., ubiquitin gene) are superior to the CaMV 35S promoter in driving Cas9 protein expression, as the latter is more susceptible to transgene silencing in vivo.[124] In legumes, both the 35S promoter[112,114,115] and the soybean EF1A2 promoter[113] have been used to achieve similar editing efficiencies.

For SNF studies, root- or nodule-specific promoters are alternative choices for driving *Cas9* gene expression during hairy root transformation, a rapid method to assay mutant phenotypes of nodulation-related genes. Compared with the 35S promoter, the leghemoglobin *LjLb2* promoter conferred similar genome-editing efficiency in transgenic hairy roots in *L. japonicus*.[114] Since nodule organogenesis involves both epidermal and cortex signaling events, it remains of great interest to test if epidermal or cortical cell-specific promoters could be exploited to investigate the biological roles of particular SNF genes with distinctive spatiotemporal expression patterns during nodulation.

The nonspecific, off-target activity of a particular Cas9/sgRNA complex is sometimes a significant concern in genome-editing experiments. In this regard, several measures have been developed to increase the targeting fidelity of Cas9, such as the nickase-type Cas9 which harbors a Cas9n (a D10A substitution) or a dCas9-*Fok*I chimera that utilizes a pair of sgRNAs targeting two neighboring sites (about 40 bp apart) on opposite DNA strands.[125–127] Also of note are novel strategies developed for human cells[128,129] that create specific structural modifications in Cas9 that increase its affinity for the target DNA sequence. These strategies have achieved higher specificity without detectable genome-wide off-target effects.

4.5 Transformation Methods for SNF Study

CRISPR/Cas9-mediated plant genome editing usually employs one of two different transformation methods, protoplast transformation or stable transformation. For legumes, hairy root transformation is an alternative method to study genes related to root biology such as SNF. Both protoplast transformation and the hairy root transformation system provide easy and fast platforms for validation of Cas9/sgRNA cutting efficiency. Protoplast transformation has been widely used in many plant systems for CRISPR-mediated gene editing experiments, including *Arabidopsis*,[130] rice,[122] wheat,[131] soybean,[112] etc.

Hairy root transformation has proven to be an effective, easily performed and time-saving procedure for CRISPR/Cas9 experiments in legumes. Sequencing of the PCR products obtained using genomic DNA extracted from transgenic hairy roots should be able to reveal the types of mutations produced and the editing efficiency supported by specific sgRNAs. Simultaneously, SNF-related phenotypes such as number of infection threads and nodules, nodule developmental defects, etc., can also be analyzed. More importantly, because hairy root transformation is a transient genome-editing method affecting only root tissues, the aboveground tissues retain their wild-type genotype. This provides a significant advantage for characterization of SNF-associated gene functions (Fig. 1).

Acquisition of heritable and stable transgenic mutant lines is essential for thorough investigations of gene functions in any field, with SNF research no exemption. Unfortunately, compared with *Arabidopsis* and rice, legumes such as soybean have a relatively low transformation efficiency with *A. tumefaciens* (~5% with soybean) and require approximately 9 months to obtain transgenic seeds.[132] Since the first application of CRISPR-mediated genome editing in soybean,[133] progress has been made with other legumes using stable transformation methods. These legumes include *L. japonicus*[114] and *M. truncatula*.[115] Biolistic transformation is an alternative method for producing transgenic soybean with a higher frequency (~9%) of regeneration and has also been used to achieve targeted gene disruption and gene replacement.[133] However, the high cost of the high-pressure helium

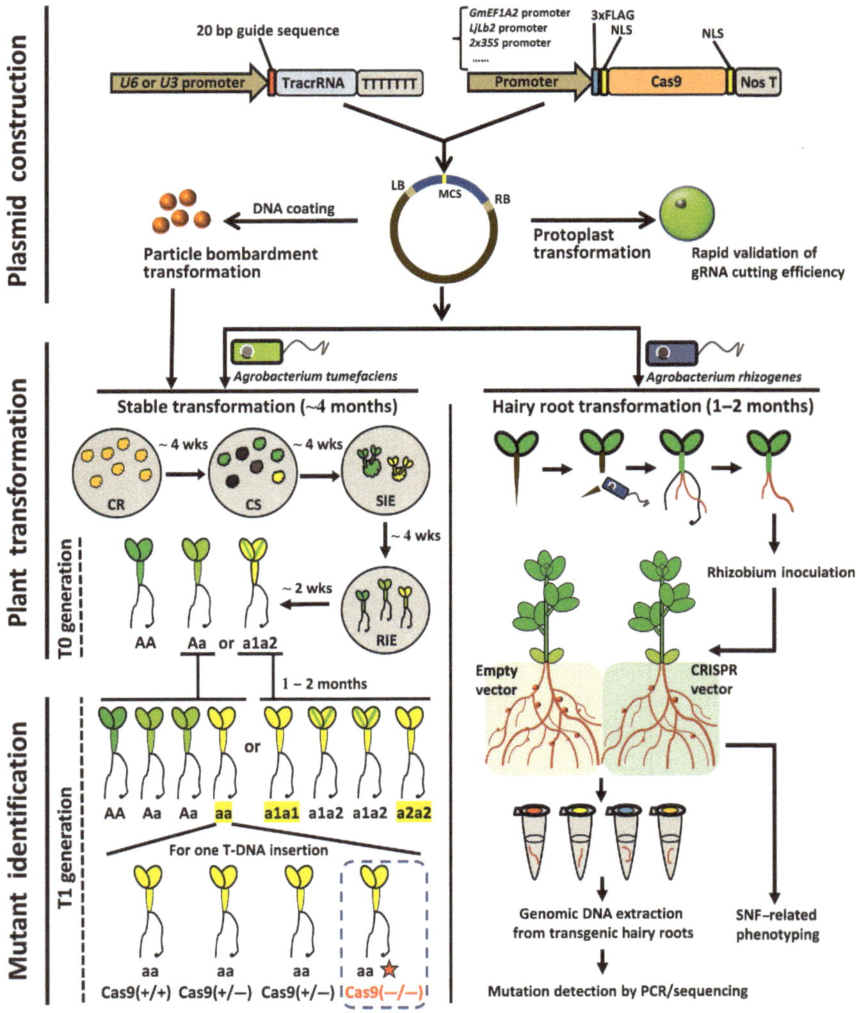

Fig. 1 Workflow for CRISPR–Cas9-mediated genome editing in legumes. The customized Cas9 and sgRNA expression units are cloned into the binary vector for particle bombardment transformation, protoplast transformation, or *Agrobacterium*-mediated stable and hairy root transformation. Using *Lotus japonicus* as an example, for *Agrobacterium tumefaciens*-mediated stable transformation, it takes around 5–6 months to obtain homozygous mutants in T1 generation via several steps of tissue culture, including callus regeneration (CR), callus selection (CS), shoot induction and elongation (SIE), and root induction and elongation (RIE). Cas9-free homozygous mutant lines could be recovered in T1 generation by self-pollination of the T0 plants and analyses of the progenies. For *Agrobacterium rhizogenes*-mediated hairy root transformation, positive transgenic roots are identified by fluorescence observation and hairy roots are formed within 1 month. Genomic DNA of transgenic hairy roots is extracted for PCR confirmation of the DNA modifications. The hairy root transformed plants can be directly analyzed for the SNF-related phenotypes. *AA*, wild type; *Aa*, *a1a2*, heterozygote; *aa*, *a1a1*, *a2a2*, homozygote; *Cas9(−/−)*, Cas9-free plant; *wks*, weeks.

microparticle acceleration system has limited its widespread use in many research laboratories.

Regardless of what procedures are used to generate transgenic plants, it is necessary to isolate marker-free transgenic lines with desired mutations. The commonly used method is to obtain progenies by self-pollination for two to three generations and to genotype the targeted mutations using site-specific PCR and DNA sequencing in each generation.[134] As legumes are strictly self-pollinating plants, this method can be adopted as a routine screening strategy. Alternatively, a visual screen for Cas9-free plants can be performed by including a gene encoding a fluorescent marker (e.g., mCherry or GFP) in the transformation vectors.[135] Recently, Zhang et al. reported an efficient method to regenerate calli that only transiently express CRISPR/Cas9 DNA when delivered by particle bombardment.[136] This experimental design is suitable for use with transformation-recalcitrant species such as soybean and woody plants.

5. PERSPECTIVES AND CONCLUDING REMARKS

Gene knock-in, gene replacement and gene-specific point modifications are all of great value for studies of functional genomics and for plant breeding. Earlier efforts to obtain transgenic knock-in plants mostly relied on the HR repair mechanism in the presence of the donor DNA.[137,138] Recently, Li *et al.*[139] utilized two sgRNAs targeting adjacent introns and the flanking sequences of a donor DNA template to accomplish precise DNA replacement in an NHEJ-dependent manner.[140] Several recent studies also demonstrated the use of a modified CRISPR system to accomplish targeted point mutations in human cells[140] and rice.[141,142] These technological improvements are expected to be applied in legumes in the near future and should be able to dramatically accelerate SNF research progress.

Although the use of CRISPR/Cas9 technology to construct targeted gene mutations is becoming a routine practice, disruption of housekeeping genes or essential genes might result in a lethal phenotype. Under these circumstances, a gene knockdown strategy known as CRISPRi (CRISPR interference) can be exploited to study the biological functions of these genes. The CRISPRi platform consists of nuclease-dead Cas9 (dCas9) and sgRNAs to direct the binding of dCas9–sgRNA to a specific promoter region and block the transcriptional initiation.[143] In contrast, a CRISPR-based transcriptional activation (CRISPRa) system can be constructed by fusing dCas9 with a transcriptional activation element such as VP64 motif.

Either CRISPRi or CRISPRa requires several sgRNAs to achieve robust suppression or activation of target gene expression. So far, these two applications have been successfully tested in *Nicotiana benthamiana* using a transient expression system.[144]

The CRISPR/Cas9 system also provides a novel strategy for forward genetic screens. Recently, several reports demonstrated utilization of CRISPR combining genomic-scale sgRNA libraries for genetic screening in human or mammalian cells.[145–147] We propose that development of a genomic-scale legume sgRNAs library and the CRISPR/Cas9 gene-editing technique could provide a mass of novel and significant insights into the nodulation signaling mechanisms.

The past few years have witnessed an explosive increase in the applications of the CRISPR/Cas technology in many model plants and crops including legumes, largely due to the simplicity and affordability of this technology. The CRISPR/Cas9 system provides new avenues for legumes functional genomics studies in general, and SNF in particular. Advances allowed by these studies will assist molecular breeding and may provide genetically modified legumes or nonlegume crops with improved nitrogen use efficiency (NUE), one of the main objectives of modern agriculture.

ACKNOWLEDGMENTS

The research in our laboratory was supported by National Natural Science Foundation of China (Project No. 31570233), the Fundamental Research Funds for the Central Universities (Program No. 2662015PY171), and Huazhong Agricultural University Scientific & Technological Self-innovation Foundation (Program No. 2014RC018).

REFERENCES

1. Farmer B. Perspectives on the 'green revolution' in South Asia. *Mod Asian Stud.* 1986;20(01):175–199.
2. Tester M, Langridge P. Breeding technologies to increase crop production in a changing world. *Science.* 2010;327(5967):818–822.
3. Fowler D, Coyle M, Skiba U, et al. The global nitrogen cycle in the twenty-first century. *Philos Trans R Soc Lond B Biol Sci.* 2013;368(1621):20130164.
4. Oldroyd GE, Murray JD, Poole PS, Downie JA. The rules of engagement in the legume-rhizobial symbiosis. *Annu Rev Genet.* 2011;45:119–144.
5. Graham PH, Vance CP. Legumes: importance and constraints to greater use. *Plant Physiol.* 2003;131(3):872–877.
6. Sato S, Nakamura Y, Kaneko T, et al. Genome structure of the legume, Lotus japonicus. *DNA Res.* 2008;15(4):227–239.
7. Young ND, Debelle F, Oldroyd GE, et al. The Medicago genome provides insight into the evolution of rhizobial symbioses. *Nature.* 2011;480(7378):520–524.
8. Schmutz J, Cannon SB, Schlueter J, et al. Genome sequence of the palaeopolyploid soybean. *Nature.* 2010;463(7278):178–183.

9. Mus F, Crook MB, Garcia K, et al. Symbiotic nitrogen fixation and the challenges to its extension to nonlegumes. *Appl Environ Microbiol.* 2016;82(13):3698–3710.
10. Oldroyd GE, Downie JA. Coordinating nodule morphogenesis with rhizobial infection in legumes. *Annu Rev Plant Biol.* 2008;59:519–546.
11. Radutoiu S, Madsen LH, Madsen EB, et al. Plant recognition of symbiotic bacteria requires two LysM receptor-like kinases. *Nature.* 2003;425(6958):585–592.
12. Limpens E, Franken C, Smit P, Willemse J, Bisseling T, Geurts R. LysM domain receptor kinases regulating rhizobial Nod factor-induced infection. *Science.* 2003; 302(5645):630–633.
13. Smit P, Limpens E, Geurts R, et al. Medicago LYK3, an entry receptor in rhizobial nodulation factor signaling. *Plant Physiol.* 2007;145(1):183–191.
14. Indrasumunar A, Searle I, Lin MH, et al. Nodulation factor receptor kinase 1alpha controls nodule organ number in soybean (*Glycine max* L. Merr). *Plant J.* 2011;65(1):39–50.
15. Madsen EB, Madsen LH, Radutoiu S, et al. A receptor kinase gene of the LysM type is involved in legume perception of rhizobial signals. *Nature.* 2003;425(6958):637–640.
16. Indrasumunar A, Kereszt A, Searle I, et al. Inactivation of duplicated nod factor receptor 5 (NFR5) genes in recessive loss-of-function non-nodulation mutants of allotetraploid soybean (*Glycine max* L. Merr.). *Plant Cell Physiol.* 2010;51(2):201–214.
17. Arrighi JF, Barre A, Ben Amor B, et al. The *Medicago truncatula* lysine motif-receptor-like kinase gene family includes NFP and new nodule-expressed genes. *Plant Physiol.* 2006;142(1):265–279.
18. Kawaharada Y, Kelly S, Nielsen MW, et al. Receptor-mediated exopolysaccharide perception controls bacterial infection. *Nature.* 2015;523(7560):308–312.
19. Stracke S, Kistner C, Yoshida S, et al. A plant receptor-like kinase required for both bacterial and fungal symbiosis. *Nature.* 2002;417(6892):959–962.
20. Endre G, Kereszt A, Kevei Z, Mihacea S, Kalo P, Kiss GB. A receptor kinase gene regulating symbiotic nodule development. *Nature.* 2002;417(6892):962–966.
21. Indrasumunar A, Wilde J, Hayashi S, Li D, Gresshoff PM. Functional analysis of duplicated Symbiosis receptor kinase (SymRK) genes during nodulation and mycorrhizal infection in soybean (*Glycine max*). *J Plant Physiol.* 2015;176(9):157–168.
22. Anè JM, Kiss GB, Riely BK, et al. *Medicago truncatula* DMI1 required for bacterial and fungal symbioses in legumes. *Science.* 2004;303(5662):1364–1367.
23. Imaizumi-Anraku H, Takeda N, Charpentier M, et al. Plastid proteins crucial for symbiotic fungal and bacterial entry into plant roots *Nature.* 2005;433(7025):527–531.
24. Charpentier M, Sun J, Martins TV, et al. Nuclear-localized cyclic nucleotide-gated channels mediate symbiotic calcium oscillations. *Science.* 2016;352(6289):1102–1105.
25. Capoen W, Sun J, Wysham D, et al. Nuclear membranes control symbiotic calcium signaling of legumes. *Proc Natl Acad Sci USA..* 2011;108(34):14348–14353.
26. Levy J, Bres C, Geurts R, et al. A putative Ca^{2+} and calmodulin-dependent protein kinase required for bacterial and fungal symbioses. *Science.* 2004;303(5662):1361–1364.
27. Tirichine L, Imaizumi-Anraku H, Yoshida S, et al. Deregulation of a Ca^{2+}/calmodulin-dependent kinase leads to spontaneous nodule development. *Nature.* 2006; 441(7097):1153–1156.
28. Messinese E, Mun JH, Yeun LH, et al. A novel nuclear protein interacts with the symbiotic DMI3 calcium- and calmodulin-dependent protein kinase of *Medicago truncatula. Mol Plant Microbe Interact.* 2007;20(8):912–921.
29. Yano K, Yoshida S, Muller J, et al. CYCLOPS, a mediator of symbiotic intracellular accommodation. *Proc Natl Acad Sci USA.* 2008;105(51):20540–20545.
30. Schauser L, Roussis A, Stiller J, Stougaard J. A plant regulator controlling development of symbiotic root nodules. *Nature.* 1999;402(6758):191–195.
31. Marsh JF, Rakocevic A, Mitra RM, et al. *Medicago truncatula* NIN is essential for rhizobial-independent nodule organogenesis induced by autoactive calcium/calmodulin-dependent protein kinase. *Plant Physiol.* 2007;144(1):324–335.

32. Hayashi S, Reid DE, Lorenc MT, et al. Transient Nod factor-dependent gene expression in the nodulation-competent zone of soybean (*Glycine max* [L.] Merr.) roots. *Plant Biotechnol J*. 2012;10(8):995–1010.
33. Heckmann AB, Lombardo F, Miwa H, et al. *Lotus japonicus* nodulation requires two GRAS domain regulators, one of which is functionally conserved in a non-legume. *Plant Physiol*. 2006;142(4):1739–1750.
34. Smit P, Raedts J, Portyanko V, et al. NSP1 of the GRAS protein family is essential for rhizobial Nod factor-induced transcription. *Science*. 2005;308(5729):1789–1791.
35. Kaló P, Gleason C, Edwards A, et al. Nodulation signaling in legumes requires NSP2, a member of the GRAS family of transcriptional regulators. *Science*. 2005; 308(5729):1786–1789.
36. Middleton PH, Jakab J, Penmetsa RV, et al. An ERF transcription factor in *Medicago truncatula* that is essential for nod factor signal transduction. *Plant Cell*. 2007;19(4):1221–1234.
37. Yano K, Aoki S, Liu M, et al. Function and evolution of a *Lotus japonicus* AP2/ERF family transcription factor that is required for development of infection threads. *DNA Res*. 2016. pii: dsw052. http://dx.doi.org/10.1093/dnares/dsw052.
38. Gonzalez-Rizzo S, Crespi M, Frugier F. The *Medicago truncatula* CRE1 cytokinin receptor regulates lateral root development and early symbiotic interaction with Sinorhizobium meliloti. *Plant Cell*. 2006;18(10):2680–2693.
39. Murray JD, Karas BJ, Sato S, Tabata S, Amyot L, Szczyglowski K. A cytokinin perception mutant colonized by Rhizobium in the absence of nodule organogenesis. *Science*. 2007;315(5808):101–104.
40. Tirichine L, Sandal N, Madsen LH, et al. A gain-of-function mutation in a cytokinin receptor triggers spontaneous root nodule organogenesis. *Science*. 2007; 315(5808):104–107.
41. Combier JP, Frugier F, de Billy F, et al. MtHAP2-1 is a key transcriptional regulator of symbiotic nodule development regulated by micro RNA169 in *Medicago truncatula*. *Genes Dev*. 2006;20(22):3084–3088.
42. Schaarschmidt S, Gresshoff PM, Hause B. Analyzing the soybean transcriptome during autoregulation of mycorrhization identifies the transcription factors GmNF-YA1a/b as positive regulators of arbuscular mycorrhization. *Genome Biol*. 2013;14(6):R62.
43. Soyano T, Kouchi H, Hirota A, Hayashi M. Nodule inception directly targets NF-Y subunit genes to regulate essential processes of root nodule development in *Lotus japonicus*. *PLoS Genet*. 2013;9(3):e1003352.
44. Nishimura R, Hayashi M, Wu GJ, et al. HAR1 mediates systemic regulation of symbiotic organ development. *Nature*. 2002;420(6914):426–429.
45. Krusell L, Madsen LH, Sato S, et al. Shoot control of root development and nodulation is mediated by a receptor-like kinase. *Nature*. 2002;420(6914):422–426.
46. Searle IR, Men AE, Laniya TS, et al. Long-distance signaling in nodulation directed by a CLAVATA1-like receptor kinase. *Science*. 2003;299(5603):109–112.
47. Schnabel E, Journet EP, de Carvalho-Niebel F, Duc G, Frugoli J. The *Medicago truncatula* SUNN gene encodes a CLV1-like leucine-rich repeat receptor kinase that regulates nodule number and root length. *Plant Mol Biol*. 2005;58(6):809–822.
48. Okamoto S, Shinohara H, Mori T, Matsubayashi Y, Kawaguchi M. Root-derived CLE glycopeptides control nodulation by direct binding to HAR1 receptor kinase. *Nat Commun*. 2013;4(4):2191.
49. Mortier V, Den Herder G, Whitford R, et al. CLE peptides control *Medicago truncatula* nodulation locally and systemically. *Plant Physiol*. 2010;153(1):222–237.
50. Lim CW, Lee YW, Hwang CH. Soybean nodule-enhanced CLE peptides in roots act as signals in GmNARK-mediated nodulation suppression. *Plant Cell Physiol*. 2011; 52(9):1613–1627.

51. Reid DE, Ferguson BJ, Gresshoff PM. Inoculation- and nitrate-induced CLE peptides of soybean control NARK-dependent nodule formation. *Mol Plant Microbe Interact.* 2011;24(5):606–618.
52. Magori S, Oka-Kira E, Shibata S, et al. Too much love, a root regulator associated with the long-distance control of nodulation in *Lotus japonicus*. *Mol Plant Microbe Interact.* 2009;22(3):259–268.
53. Sprent JI, James EK. Legume evolution: where do nodules and mycorrhizas fit in? *Plant Physiol.* 2007;144(2):575–581.
53a. Okazaki S, Kaneko T, Sato S, Saeki K. Hijacking of leguminous nodulation signaling by the rhizobial type III secretion system. *Proc Natl Acad Sci USA.* 2013;110 (42):17131–17136.
54. Broghammer A, Krusell L, Blaise M, et al. Legume receptors perceive the rhizobial lipochitin oligosaccharide signal molecules by direct binding. *Proc Natl Acad Sci USA.* 2012;109(34):13859–13864.
55. Venkateshwaran M, Jayaraman D, Chabaud M, et al. A role for the mevalonate pathway in early plant symbiotic signaling. *Proc Natl Acad Sci USA.* 2015;112(31):9781–9786.
56. Gleason C, Chaudhuri S, Yang T, Munoz A, Poovaiah BW, Oldroyd GE. Nodulation independent of rhizobia induced by a calcium-activated kinase lacking autoinhibition. *Nature.* 2006;441(7097):1149–1152.
57. Miller JB, Pratap A, Miyahara A, et al. Calcium/calmodulin-dependent protein kinase is negatively and positively regulated by calcium, providing a mechanism for decoding calcium responses during symbiosis signaling. *Plant Cell.* 2013;25(12):5053–5066.
58. Singh S, Katzer K, Lambert J, Cerri M, Parniske M. CYCLOPS, a DNA-binding transcriptional activator, orchestrates symbiotic root nodule development. *Cell Host Microbe.* 2014;15(2):139–152.
59. Murakami Y, Miwa H, Imaizumi-Anraku H, et al. Positional cloning identifies *Lotus japonicus* NSP2, a putative transcription factor of the GRAS family, required for NIN and ENOD40 gene expression in nodule initiation. *DNA Res.* 2006;13(6):255–265.
60. Xie F, Murray JD, Kim J, et al. Legume pectate lyase required for root infection by rhizobia. *Proc Natl Acad Sci USA.* 2012;109(2):633–638.
61. Soyano T, Hirakawa H, Sato S, Hayashi M, Kawaguchi M. Nodule inception creates a long-distance negative feedback loop involved in homeostatic regulation of nodule organ production. *Proc Natl Acad Sci USA.* 2014;111(40):14607–14612.
62. Vernie T, Kim J, Frances L, et al. The NIN transcription factor coordinates diverse nodulation programs in different tissues of the *Medicago truncatula* root. *Plant Cell.* 2015;27(12):3410–3424.
63. Ryu H, Cho H, Choi D, Hwang I. Plant hormonal regulation of nitrogen-fixing nodule organogenesis. *Mol Cells.* 2012;34(2):117–126.
64. Held M, Hou H, Miri M, et al. *Lotus japonicus* cytokinin receptors work partially redundantly to mediate nodule formation. *Plant Cell.* 2014;26(2):678–694.
65. Moubayidin L, Di Mambro R, Sabatini S. Cytokinin-auxin crosstalk. *Trends Plant Sci.* 2009;14(10):557–562.
66. Wang Y, Wang L, Zou Y, et al. Soybean miR172c targets the repressive AP2 transcription factor NNC1 to activate ENOD40 expression and regulate nodule initiation. *Plant Cell.* 2014;26(12):4782–4801.
67. Suzuki A, Suriyagoda L, Shigeyama T, et al. *Lotus japonicus* nodulation is photomorphogenetically controlled by sensing the red/far red (R/FR) ratio through jasmonic acid (JA) signaling. *Proc Natl Acad Sci USA.* 2011;108(40):16837–16842
68. Mortier V, Holsters M, Goormachtig S. Never too many? How legumes control nodule numbers. *Plant Cell Environ.* 2012;35(2):245–258.
69. Kryvoruchko IS, Sinharoy S, Torres-Jerez I, et al. MtSWEET11, a nodule-specific sucrose transporter of *Medicago truncatula* root nodules. *Plant Physiol.* 2016; 171(1):554–565.

70. Vance CP, Gantt JS. Control of nitrogen and carbon metabolism in root nodules. *Physiol Plant.* 1992;85(2):266–274.
71. Krusell L, Krause K, Ott T, et al. The sulfate transporter SST1 is crucial for symbiotic nitrogen fixation in *Lotus japonicus* root nodules. *Plant Cell.* 2005;17(5):1625–1636.
72. Takanashi K, Yokosho K, Saeki K, et al. LjMATE1: a citrate transporter responsible for iron supply to the nodule infection zone of *Lotus japonicus. Plant Cell Physiol.* 2013;54(4):585–594.
73. Udvardi M, Poole PS. Transport and metabolism in legume-rhizobia symbioses. *Annu Rev Plant Biol.* 2013;64:781–805.
74. Ott T, van Dongen JT, Gunther C, et al. Symbiotic leghemoglobins are crucial for nitrogen fixation in legume root nodules but not for general plant growth and development. *Curr Biol.* 2005;15(6):531–535.
75. Puppo A, Groten K, Bastian F, et al. Legume nodule senescence: roles for redox and hormone signalling in the orchestration of the natural aging process. *New Phytol.* 2005;165(3):683–701.
76. Loscos J, Matamoros MA, Becana M. Ascorbate and homoglutathione metabolism in common bean nodules under stress conditions and during natural senescence. *Plant Physiol.* 2008;146(3):1282–1292.
77. van Wyk SG, Du Plessis M, Cullis CA, Kunert KJ, Vorster BJ. Cysteine protease and cystatin expression and activity during soybean nodule development and senescence. *BMC Plant Biol.* 2014;14:294.
78. Li Y, Zhou L, Li Y, et al. A nodule-specific plant cysteine proteinase, AsNODF32, is involved in nodule senescence and nitrogen fixation activity of the green manure legume Astragalus sinicus. *New Phytol.* 2008;180(1):185–192.
79. Yuan S, Li R, Wang L, et al. Search for nodulation and nodule development-related cystatin genes in the genome of soybean (*Glycine max*). *Front Plant Sci.* 2016;7:1595.
80. Bolon YT, Haun WJ, Xu WW, et al. Phenotypic and genomic analyses of a fast neutron mutant population resource in soybean. *Plant Physiol.* 2011;156(1):240–253.
81. Cooper JL, Till BJ, Laport RG, et al. TILLING to detect induced mutations in soybean. *BMC Plant Biol.* 2008;8:9.
82. Cui Y, Barampuram S, Stacey MG, et al. *Tnt1* retrotransposon mutagenesis: a tool for soybean functional genomics. *Plant Physiol.* 2013;161(1):36–47.
83. Malolepszy A, Mun T, Sandal N, et al. The *LORE1* insertion mutant resource. *Plant J.* 2016;88(2):306–317.
84. Perry J, Brachmann A, Welham T, et al. TILLING in *Lotus japonicus* identified large allelic series for symbiosis genes and revealed a bias in functionally defective ethyl methanesulfonate alleles toward glycine replacements. *Plant Physiol.* 2009; 151(3):1281–1291.
85. Iantcheva A, Revalska M, Zehirov G, et al. *Tnt1* retrotransposon as an efficient tool for development of an insertional mutant collection of *Lotus japonicus. In Vitro Cell Dev Biol Plant.* 2016;52(3):338–347.
86. Tadege M, Wen J, He J, et al. Large-scale insertional mutagenesis using the *Tnt1* retrotransposon in the model legume *Medicago truncatula. Plant J.* 2008;54(2):335–347.
87. Le Signor C, Savois V, Aubert G, et al. Optimizing TILLING populations for reverse genetics in *Medicago truncatula. Plant Biotechnol J.* 2009;7(5):430–441.
88. Kawaguchi M, Imaizumi-Anraku H, Koiwa H, et al. Root, root hair, and symbiotic mutants of the model legume *Lotus japonicus. Mol Plant Microbe Interact.* 2002; 15(1):17–26.
89. Perry JA, Wang TL, Welham TJ, et al. A TILLING reverse genetics tool and a web-accessible collection of mutants of the legume *Lotus japonicus. Plant Physiol.* 2003;131(3):866–871.

90. d'Erfurth I, Cosson V, Eschstruth A, Lucas H, Kondorosi A, Ratet P. Efficient transposition of the *Tnt1* tobacco retrotransposon in the model legume *Medicago truncatula*. *Plant J*. 2003;34(1):95–106.
91. Małolepszy A, Mun T, Sandal N, et al. The *LORE1* insertion mutant resource. *Plant J*. 2016;88(2):306–317.
92. Urbański DF, Małolepszy A, Stougaard J, Andersen SU. Genome-wide *LORE1* retrotransposon mutagenesis and high-throughput insertion detection in *Lotus japonicus*. *Plant J*. 2012;69(4):731–741.
93. Fukai E, Soyano T, Umehara Y, et al. Establishment of a *Lotus japonicus* gene tagging population using the exon-targeting endogenous retrotransposon *LORE1*. *Plant J*. 2012;69(4):720–730.
94. Chaintreuil C, Rivallan R, Bertioli DJ, et al. A gene-based map of the nod factor-independent Aeschynomene evenia sheds new light on the evolution of nodulation and legume genomes. *DNA Res*. 2016;23(4):365–376.
95. Urnov FD, Rebar EJ, Holmes MC, Zhang HS, Gregory PD. Genome editing with engineered zinc finger nucleases. *Nat Rev Genet*. 2010;11(9):636–646.
96. Boch J, Scholze H, Schornack S, et al. Breaking the code of DNA binding specificity of TAL-type III effectors. *Science*. 2009;326(5959):1509–1512.
97. Miller JC, Tan S, Qiao G, et al. A TALE nuclease architecture for efficient genome editing. *Nat Biotechnol*. 2011;29(2):143–148.
98. Curtin SJ, Zhang F, Sander JD, et al. Targeted mutagenesis of duplicated genes in soybean with zinc-finger nucleases. *Plant Physiol*. 2011;156(2):466–473.
99. Haun W, Coffman A, Clasen BM, et al. Improved soybean oil quality by targeted mutagenesis of the fatty acid desaturase 2 gene family. *Plant Biotechnol J*. 2014; 12(7):934–940.
100. Gaj T, Gersbach CA, Barbas 3rd CF. ZFN, TALEN, and CRISPR/Cas-based methods for genome engineering. *Trends Biotechnol*. 2013;31(7):397–405.
101. Barrangou R, Marraffini LA. CRISPR-Cas systems: prokaryotes upgrade to adaptive immunity. *Mol Cell*. 2014;54(2):234–244.
102. Horvath P, Barrangou R. CRISPR/Cas, the immune system of bacteria and archaea. *Science*. 2010;327(5962):167–170.
103. Makarova KS, Haft DH, Barrangou R, et al. Evolution and classification of the CRISPR-Cas systems. *Nat Rev Microbiol*. 2011;9(6):467–477.
104. Jinek M, Chylinski K, Fonfara I, Hauer M, Doudna JA, Charpentier E. A programmable dual-RNA-guided DNA endonuclease in adaptive bacterial immunity. *Science*. 2012;337(6096):816–821.
105. Cong L, Ran FA, Cox D, et al. Multiplex genome engineering using CRISPR/Cas systems. *Science*. 2013;339(6121):819–823.
106. Ding Y, Li H, Chen LL, Xie K. Recent advances in genome editing using CRISPR/Cas9. *Front Plant Sci*. 2016;7:703.
107. Li JF, Norville JE, Aach J, et al. Multiplex and homologous recombination-mediated genome editing in Arabidopsis and *Nicotiana benthamiana* using guide RNA and Cas9. *Nat Biotechnol*. 2013;31(8):688–691.
108. Feng Z, Mao Y, Xu N, et al. Multigeneration analysis reveals the inheritance, specificity, and patterns of CRISPR/Cas-induced gene modifications in Arabidopsis. *Proc Natl Acad Sci USA*. 2014;111(12):4632–4637.
109. Shan QW, Wang YP, Li J, et al. Targeted genome modification of crop plants using a CRISPR-Cas system. *Nat Biotechnol*. 2013;31(8):686–688.
110. Brooks C, Nekrasov V, Lippman ZB, Van Eck J. Efficient gene editing in tomato in the first generation using the clustered regularly interspaced short palindromic repeats/CRISPR-associated 9 system. *Plant Physiol*. 2014;166(3):1292–1297.

111. Wang S, Zhang S, Wang W, Xiong X, Meng F, Cui X. Efficient targeted mutagenesis in potato by the CRISPR/Cas9 system. *Plant Cell Rep*. 2015;34(9):1473–1476.
112. Sun X, Hu Z, Chen R, et al. Targeted mutagenesis in soybean using the CRISPR-Cas9 system. *Sci Rep*. 2015;5:10342.
113. Li Z, Liu ZB, Xing A, et al. Cas9-guide RNA directed genome editing in soybean. *Plant Physiol*. 2015;169(2):960–970.
114. Wang L, Wang L, Tan Q, et al. Efficient inactivation of symbiotic nitrogen fixation related genes in *Lotus japonicus* using CRISPR-Cas9. *Front Plant Sci*. 2016;7:1333.
115. Meng Y, Hou Y, Wang H, et al. Targeted mutagenesis by CRISPR/Cas9 system in the model legume *Medicago truncatula*. *Plant Cell Rep*. 2016;36(2):371–374.
116. Cai Y, Chen L, Liu X, et al. CRISPR/Cas9-mediated genome editing in soybean hairy roots. *PLoS One*. 2015;10(8):e0136064.
117. Jacobs TB, LaFayette PR, Schmitz RJ, Parrott WA. Targeted genome modifications in soybean with CRISPR/Cas9. *BMC Biotechnol*. 2015;15(1):1–10.
118. Du H, Zeng X, Zhao M, et al. Efficient targeted mutagenesis in soybean by TALENs and CRISPR/Cas9. *J Biotechnol*. 2016;217:90–97.
119. Michno JM, Wang X, Liu J, Curtin SJ, Kono TJ, Stupar RM. CRISPR/Cas mutagenesis of soybean and *Medicago truncatula* using a new web-tool and a modified Cas9 enzyme. *GM Crops Food*. 2015;6(4):243–252.
120. Xie K, Zhang J, Yang Y. Genome-wide prediction of highly specific guide RNA spacers for CRISPR-Cas9-mediated genome editing in model plants and major crops. *Mol Plant*. 2014;7(5):923–926.
121. Lei Y, Lu L, Liu HY, Li S, Xing F, Chen LL. CRISPR-P: a web tool for synthetic single-guide RNA design of CRISPR-system in plants. *Mol Plant*. 2014;7(9):1494–1496.
122. Xie K, Minkenberg B, Yang Y. Boosting CRISPR/Cas9 multiplex editing capability with the endogenous tRNA-processing system. *Proc Natl Acad Sci USA*. 2015;112(11):3570–3575.
123. Zhang Z, Mao Y, Ha S, Liu W, Botella JR, Zhu JK. A multiplex CRISPR/Cas9 platform for fast and efficient editing of multiple genes in Arabidopsis. *Plant Cell Rep*. 2015;35(7):1519–1533.
124. Jiang W, Yang B, Weeks DP. Efficient CRISPR/Cas9-mediated gene editing in *Arabidopsis thaliana* and inheritance of modified genes in the T2 and T3 generations. *PLoS One*. 2014;9(6):e99225.
125. Guilinger JP, Thompson DB, Liu DR. Fusion of catalytically inactive Cas9 to FokI nuclease improves the specificity of genome modification. *Nat Biotechnol*. 2014; 32(6):577–582.
126. Ran FA, Hsu PD, Lin CY, et al. Double nicking by RNA-guided CRISPR Cas9 for enhanced genome editing specificity. *Cell*. 2013;154(6):1380–1389.
127. Tsai SQ, Wyvekens N, Khayter C, et al. Dimeric CRISPR RNA-guided FokI nucleases for highly specific genome editing. *Nat Biotechnol*. 2014;32(6):569–576.
128. Kleinstiver BP, Pattanayak V, Prew MS, et al. High-fidelity CRISPR–Cas9 nucleases with no detectable genome-wide off-target effects. *Nature*. 2016;529(7587):490–495.
129. Slaymaker IM, Gao L, Zetsche B, Scott DA, Yan WX, Zhang F. Rationally engineered Cas9 nucleases with improved specificity. *Science*. 2016;351(6268):84–88.
130. Feng Z, Zhang B, Ding W, et al. Efficient genome editing in plants using a CRISPR/Cas system. *Cell Res*. 2013;23(10):1229–1232.
131. Shan Q, Wang Y, Li J, Gao C. Genome editing in rice and wheat using the CRISPR/Cas system. *Nat Protoc*. 2014;9(10):2395–2410.
132. Luth D, Warnberg K, Wang K. Soybean [*Glycine max* (L.) Merr]. *Methods Mol Biol*. 2015;1223:275–284.

133. Rech EL, Vianna GR, Aragao FJ. High-efficiency transformation by biolistics of soybean, common bean and cotton transgenic plants. *Nat Protoc.* 2008;3(3):410–418.
134. Xu RF, Li H, Qin RY, et al. Generation of inheritable and "transgene clean" targeted genome-modified rice in later generations using the CRISPR/Cas9 system. *Sci Rep.* 2015;5:11491.
135. Gao X, Chen J, Dai X, Zhang D, Zhao Y. An effective strategy for reliably isolating heritable and Cas9-free Arabidopsis mutants generated by CRISPR/Cas9-mediated genome editing. *Plant Physiol.* 2016;171(3):1794–1800.
136. Zhang Y, Liang Z, Zong Y, et al. Efficient and transgene-free genome editing in wheat through transient expression of CRISPR/Cas9 DNA or RNA. *Nat Commun.* 2016;7:12617.
137. Sun Y, Zhang X, Wu C, et al. Engineering herbicide-resistant rice plants through CRISPR/Cas9-mediated homologous recombination of acetolactate synthase. *Mol Plant.* 2016;9(4):628–631.
138. Cermak T, Baltes NJ, Cegan R, Zhang Y, Voytas DF. High-frequency, precise modification of the tomato genome. *Genome Biol.* 2015;16:232.
139. Li J, Meng X, Zong Y, et al. Gene replacements and insertions in rice by intron targeting using CRISPR-Cas9. *Nat Plants.* 2016;2:16139.
140. Komor AC, Kim YB, Packer MS, Zuris JA, Liu DR. Programmable editing of a target base in genomic DNA without double-stranded DNA cleavage. *Nature.* 2016; 533(7603):420–424.
141. Li J, Sun Y, Du J, Zhao Y, Xia L. Generation of targeted point mutations in rice by a modified CRISPR/Cas9 system. *Mol Plant.* 2016;10(3):526–529.
142. Lu Y, Zhu JK. Precise editing of a target base in the rice genome using a modified CRISPR/Cas9 system. *Mol Plant.* 2016;10(3):523–525.
143. Larson MH, Gilbert LA, Wang X, Lim WA, Weissman JS, Qi LS. CRISPR interference (CRISPRi) for sequence-specific control of gene expression. *Nat Protoc.* 2013;8(11):2180–2196.
144. Piatek A, Ali Z, Baazim H, et al. RNA-guided transcriptional regulation in planta via synthetic dCas9-based transcription factors. *Plant Biotechnol J.* 2015;13(4):578–589.
145. Gilbert LA, Horlbeck MA, Adamson B, et al. Genome-scale CRISPR-mediated control of gene repression and activation. *Cell.* 2014;159(3):647–661.
146. Wang T, Wei JJ, Sabatini DM, Lander ES. Genetic screens in human cells using the CRISPR-Cas9 system. *Science.* 2014;343(6166):80–84.
147. Shalem O, Sanjana NE, Hartenian E, et al. Genome-scale CRISPR-Cas9 knockout screening in human cells. *Science.* 2014;343(6166):84–87.
148. Tang X, Zheng X, Qi Y, et al. A single transcript CRISPR-Cas9 system for efficient genome editing in plants. *Mol Plant.* 2016;9(7):1088–1091.

CHAPTER TWELVE

Safety, Security, and Policy Considerations for Plant Genome Editing

Jeffrey D. Wolt[1]
Biosafety Institute for Genetically Modified Agricultural Products, Iowa State University, Ames, IA, United States
Crop Bioengineering Consortium, Iowa State University, Ames, IA, United States
[1]Corresponding author: e-mail address: jdwolt@iastate.edu

Contents

Abstract

Genome editing with engineered nucleases (GEEN) is increasingly used as a tool for gene discovery and trait development in crops through generation of targeted changes in endogenous genes. The development of the CRISPR–Cas9 system (clustered regularly interspaced short palindromic repeats with associated Cas9 protein), in particular, has enabled widespread use of genome editing. Research to date has not comprehensively addressed genome-editing specificity and off-target mismatches that may result in unintended changes within plant genomes or the potential for gene drive initiation. Governance and regulatory considerations for bioengineered crops derived from using

Progress in Molecular Biology and Translational Science, Volume 149
ISSN 1877-1173
http://dx.doi.org/10.1016/bs.pmbts.2017.03.005

GEEN will require greater clarity as to target specificity, the potential for mismatched edits, unanticipated downstream effects of off-target mutations, and assurance that genome reagents do not occur in finished products. Since governance and regulatory decision making involves robust standards of evidence extending from the laboratory to the postcommercial marketplace, developers of genome-edited crops must anticipate significant engagement and investment to address questions of regulators and civil society.

1. INTRODUCTION

Fundamental and applied advances to understand plant gene function and to generate useful crop traits are being facilitated through the emergence and application of genome-editing techniques. These techniques allow for site-directed edits that result in simple point mutations involving insertions/deletions of one to several nucleotides with subsequent repair through nonhomologous end joining (NHEJ), homology-directed repair (HDR) involving template insertions, or targeted transgene insertions. Each of these editing outcomes can be achieved through a variety of editing reagents and approaches (Box 1), but it is the emergence of the clustered regularly interspaced short palindromic repeats/Cas9 (CRISPR–Cas9) system with its flexibility and amenability for increased editing efficiency and highly specific targeting that has democratized the field of plant genome engineering.

The use of GEEN for development of commercial applications depends on clear means to address scientific evidence and public questions regarding safety and use within robust governance and regulatory processes. Here, these aspects of GEEN are reviewed and updated with respect to their implications to downstream products created using genome-editing techniques and how these are understood by civil society and regulators. The safety and security of food systems and the environment and emerging policies in terms of research, regulation, and governance of GEEN-derived products are addressed with emphasis on CRISPR–Cas9-meditated editing but with due consideration of the use of genome-editing technologies in general.

1.1 Background and State of the Science

Multiple lines of investigation have converged in the successful application and rapid progress of GEEN in microbial and animal systems.[1,2] The applications of GEEN in higher plants have been more measured, since in order to achieve the plant genome edit, complex transformation techniques are

BOX 1 Genome-Editing Reagents and Processes

Genome editing can be accomplished with a number of reagent systems; and each elicits similar questions in terms of safety and security of derived products, and the types of regulatory and governance issues that may need to be addressed as these reagents become increasingly common for deriving novel traits of interest for crop improvement.

The most widely recognized and highly developed reagent for genome editing is the CRISPR–Cas9 system of clustered regularly interspaced short palindromic repeats in conjunction with the Cas9 endonuclease from *Streptococcus pyogenes* (*Spy*Cas9). What is commonly termed CRISPR entails an ever larger suite of associated programmable nucleases derived from bacterial and archaeal sources[112–114] that may be synthesized to alter their site recognition and binding characteristics.[115] These engineered nucleases are directed to a recognition site (a target sequence terminating in a protospacer adjacent motif) with a site-specific guide RNA (gRNA) sequence where they cause double-strand breaks (DSBs) subject to repair by endogenous mechanisms. Modified forms of CRISPR act as nickases and can cause paired off-set single-strand breaks at target sites.[116]

The earliest and most commercially advanced form of plant genome editing is oligonucleotide-mediated mutagenesis (OMM, also termed oligonucleotide-directed mutagenesis). Site-specific mutation is accomplished in OMM using chemically synthesized oligonucleotides which, other than for the intended nucleotide modification, are homologous to the target site. Cleavage and repair is directed by endogenous systems resulting in low editing efficiencies.[3] Modifications of OMM can use accompanying engineered nucleases to improve editing efficiency.[117]

In addition to CRISPR, there are several additional reagent systems for genome editing with engineered nucleases (GEEN) that have been used and will continue to find future uses for specific applications. GEEN refers to any genetic engineering where DNA is inserted, replaced, or removed from a genome using a site-directed nuclease (SDN) to cause a DSB. Engineered meganucleases represent GEEN that are modified, fused, or rationally designed variants of microbially derived meganucleases.[118] These may also be referred to as LAGLIDADG endonucleases or homing nucleases. More widely used are transcriptional activator-like effector nucleases, which are programmable nucleases, comprised of the DNA binding domain of *Xanthomonas*-derived TAL effectors fused with *Fok*I restriction endonuclease[119]; and zinc-finger nucleases (ZFNs), which are programmable nucleases comprised of two domains, a sequence-specific DNA binding domain comprised of three to four tandemly arranged zinc-finger proteins and the DNA-cleaving nuclease domain of the *Fok*I restriction endonuclease.[120]

Once a gene-editing reagent has caused a DSB, the repair occurs through endogenous processes. NHEJ is a means for repair of a DSB involving an

Continued

BOX 1 Genome-Editing Reagents and Processes—cont'd

error-prone process that often causes small insertions or deletions of nucleotide bases at the DSB site resulting in mutations. HDR is a mechanism for DSB repair using a DNA sequence homologous to the break site that serves as a template for homologous recombination. Repair templates may range in size from short DNA segments to entire genes and may be native, synthetic, or transgenic.

required to recover the edit through regeneration of a whole plant rather than through transient approaches often used in microbial and animal systems. OMM was the earliest form of site-specific genome editing in higher plants where targeted base replacement or addition was facilitated using chemically synthesized oligonucleotides and endogenous DNA repair mechanisms.[3] Progress with insertion of DNA-encoded designer nucleases into model plant systems was first reported in 2009 using ZFN.[4] Subsequently, numerous proofs of concept whereby useful traits have been developed through GEEN have been described[5] involving site-specific mutations caused by NHEJ DNA repair[6] and targeted transgene insertions.[7] In animal systems, GEEN may also involve template additions through HDR, but this approach is difficult in plants where transformation involves DNA delivered to determinate cells that are not prone to HDR repair.[8] However, progress in use of CRISPR–Cas9 to stimulate HDR editing of the model plant *Arabidopsis thaliana* has been recently reported.[9]

Rapid progress in the widespread application of genome editing in plants has been achieved with the advent of systems derived from the endonuclease CRISPR–Cas9 system from *S. pyogenes*.[10,11] Improvements in targeting specificity, editing efficiencies, and the breadth of genome-editing applications are being achieved through elucidation of the properties that confer selectivity and binding of the Cas9 protein[12–14] and its associated gRNA to a complementary DNA sequence within the target gene.[15,16] Incorporation of knowledge regarding CRISPR–Cas9 design with bioinformatics tools that evaluate both match and mismatch sites will further aid efficiency and selectivity for CRISPR–Cas9 gene editing.[17]

In the area of plant genome editing, the broader adoption of GEEN for fundamental research and trait discovery research has emphasized reagent selection, nuclease source and modification, and targeting methodologies for improving the efficiency of editing of the target site.[18–20] A principal consideration has been reagent toxicity in terms of cytotoxic and genotoxic effects in targeted cells/tissue due to reagent concentration and duration of exposure (i.e., dose-dependent effects).[17] For some crops, such as soybean (*Glycine max* (L.) Merr.), reagent toxicity and resulting reduction in editing efficiency appear to limit the successful application of GEEN to somatic embryo culture.[21–23]

While there is continuing emphasis on editing efficiency for GEEN application to mammalian systems,[24–26] considerable recent attention, especially with respect to human therapeutics, has been extended to off-target edits within the genome and in the potential for downstream toxicity or other unintended effects.[27] These off-target editing effects will be less relevant to plant applications because of the ability for their detection and interdiction in the process of plant regeneration and trait segregation through breeding selection to obtain a finished product. Regardless, there is a need to improve the predictability of off-target modifications to gain wider acceptance and use of the technology for crop improvement.[17]

The introduction of CRISPR reagents (i.e., Cas9 and gRNA genes) into plant cells is typically achieved using *Agrobacterium*- or biolistic-mediated transformation methods for reagent delivery. This process leads to random insertion of the Cas9–gRNA gene cassettes into the plant genome at a site distinct from the target. Continuous genome editing to on- and off-target sites will be possible as long as the Cas9 and gRNA transgenes are present during cellular differentiation, regeneration of the whole plant, and in subsequent plant generations. These transgenic elements segregate and can be eliminated in subsequent plant breeding through selection for null segregants (NSs).[28] Dose-dependent continuous genome editing in plant tissue culture is distinct from the usual practice in animal research to transiently introduce the CRISPR reagents as Cas9 protein/gRNA ribonucleoprotein complexes or as Cas9 mRNA/gRNA mRNA into cells—processes which mitigate against continuous genome editing. Transient editing approaches are possible for plant species where protoplast (plant cells sans the cell wall) can be used for transformation and regeneration, but do not work for crops such as maize (*Zea mays*) which are difficult to regenerate from protoplasts. However, methods for direct introduction of the Cas9 or similar nucleases and gRNA into cells may lead to more widespread use of transient approaches for plant genome editing.[29,30] In vitro culture-induced mutations are an additional confounding factor in plant genome editing that may cause new phenotypes that are incorrectly ascribed to the gene-editing process (Kan Wang, personal communication).[17]

2. SAFETY AND SECURITY ASPECTS OF GENOME EDITING

As with any emerging technology, the advent of GEEN is met with public questions regarding ethics, safe use, and regulation.[31,32] Among the array of questions voiced by civil society (see Section 3), the frequency of anticipated off-target edits occurring from bioengineering and the potential for inadvertent gene drive generation are key scientific issues that will be of most interest to developing governance and regulatory direction for future products of genome editing. Answers to these questions are informed by broad research understanding developed from tools that enable precise genome editing, comprehensive genome characterization through DNA sequencing and reliable prediction of editing outcomes.

2.1 Off-Target Edits

Current evidence from CRISPR–Cas9 genome editing in plant systems indicates that the frequencies of off-target mutation are likely to be quite low. In an assessment of transgenic soybean produced during conventional *Agrobacterium*-mediated transgene insertions, genomic variation was found to be one order of magnitude less than for radiation-induced soybean mutants and two orders of magnitude less than between different soybean cultivars.[33] Establishing public trust and addressing regulatory uncertainty requires further assurances that this negligible level of off-target effects from bioengineering is indeed the general case and will also apply to Cas9/gRNA-induced mutations. Several factors enter into such analyses. For example, increased editing efficiency is observed over time for genome modifications targeted using CRISPR–Cas9 in cultured soybean embryos.[21] This may indicate continued expression of Cas9 during embryo and plant development and the potential for dose-dependent (concentration × time) effects that may increase off-target mutations. Addressing this will require further consideration of genome analysis of CRISPR-edited plant lines as well as improvements in computational tools, reagent design, and experimental methodology; collectively, these can lead to further assurances regarding the limited potential for unanticipated effects at likely mismatch sites within the genome.

2.2 Gene Drives

The likelihood for creation of gene drives is a further—and potentially more consequential—aspect of genome editing with implications for crop safety and food security. Gene drives are systems that enable biased inheritance of a genetic element so that offspring within a population have a >50% chance of inheritance of a given trait.[34] Naturally occurring gene drives are held in check due to the nature of the gene drive mechanisms, the trait they drive, the species, and environmental conditions.[34] Design of the CRISPR–Cas9 system may facilitate the creation of highly efficient gene drives when the Cas9 gene and gRNA occur within the same construct which contains flanking arms homologous to the insertion site and targets embryos or germ cells.[35] These design criteria determine the potential for the CRISPR to act as gene drive and, when present, justify staged research using clear containment/confinement conditions until potential downstream effects are better understood.[34] The creation of a gene drive may be purposeful or inadvertent, but the outcome is the same—gene expression is monoallelic as the CRISPR–Cas9 continues to target gene edits in subsequent generations.

2.2.1 Global and Local Gene Drives in Plants

The creation of gene drives in crop plants or livestock is of lesser concern than applications where gene drives are expressed in wild organisms under open release conditions—as may be the case if gene drives are used to eliminate invasive plant species or overcome pesticide resistance in weedy species.[36] For domesticated plants and animals, there is limited opportunity for uncontrolled gene drive escape and dissemination because of long generation times and control of breeding lines which, respectively, reduce gene drive efficiency and provide a means to observe and remove undesired phenotypes which may be inadvertently developed. Additionally, since domesticated crops and food animals are not competitive with sexually incompatible wild species, the probability is low for environmental establishment of gene drive-bearing crops or livestock. For similar reasons, there is limited potential that dual-use gene drive occurrence in seed supply may be a food/feed security issue. The potential for inadvertent or purposeful entry of gene drives into the seed supply is particularly low for first world applications where there is channelization of commercial seed supply. In regions of the world where farmer seed saving and farmer-to-farmer seed exchange are common, the potential for plant gene drive transmission may be higher, but long generation times and opportunities to observe and segregate undesired phenotypes remain as mitigating factors. In addition, design of gene drive reversibility, whereby a second gene drive overwrites the earlier change and restores the original function to the targeted gene, is a concept that potentially mitigates against undesired gene drive occurrence in the environment.[37]

The forgoing discussion is directed mainly to global gene drives where there is the potential for unchecked gene drive transmission through populations. Local gene drives are self-limiting systems that control the extent that a gene drive can extend monoallelic transformations through a population and, therefore, represent a design approach that further reduces the possibility for unchecked gene drive transmission. A common example of a local gene drive system is the deployment of a gene drive using daisy chains where the CRISPR-mediated mechanism for monoallelic transformation is separated in a sequence of interdependent steps.[38] A linear chain system consists of serially dependent drive elements that are unlinked and occur on separate chromosomes. The collection of elements is lost over time via natural selection. This breaks the chain to terminate monoallelic conversion, since individual elements of the chain cannot independently drive the transformation.[38]

2.3 Genome Characterization

Since the process of genome editing allows for the possibility of off-target mutagenesis (or the very rare possibility of gene drive-induced generation of monoallelic sites),[39–41] it is important to establish through molecular characterization the specificity of the genome edit that has been obtained. Molecular characterization of genetically engineered crops is used to access unintended physical effects (as well as the intended effects) of genome engineering. Genomics data (i.e., complete DNA sequencing) in conjunction with other data encompassing transcriptomic, metabolomic, epigenomic, and phenomic effects of bioengineering may be useful for inferring downstream effects which are important to efficacy and safety attributes of the bioengineered plant.[42,43] Such characterization focuses and directs subsequent risk assessment of potential effects of bioengineered crops and their derived products toward relevant questions of human and environmental safety.[43]

2.3.1 Next-Generation Sequencing

The use of PCR-based methods is useful for characterizing the site of transgene insertion in genetically engineered crops but can only partially address unanticipated off-target effects of gene edits within the genome.

Development of a variety of PCR-based genome walking methods[44] have improved the comprehensiveness of PCR-based molecular characterization but still do not address the need for genome-wide characterization of gene-editing outcomes. Improved understanding of off-target effects has led to several methods for detecting off-target edits resulting from CRISPR RNA-guided nucleases which as of yet have not been applied to plant systems.[44]

Use of advanced PCR methods for targeted deep sequencing (TDS) at near homologous off-target sites provides direct validation of off-target mutation frequencies as predicted from bioinformatics, but the failure to detect unanticipated genome-wide effects remains an uncertainty for regulatory decision making. Therefore, improved methodologies are needed to address unanticipated mutations arising from off-target gene edits.[45] A range of alternative next-generation sequencing (NGS) methods appear to be more accurate and more comprehensive for molecular characterization of gene-edited crops, but at this time suffer methodological limitations, including standardization and validation of diverse methodologies, that limit their blanket acceptance for regulatory decision making.[46] Common current alternative approaches to TDS for assessing mismatch potential for CRISPR–Cas9 include whole-genome sequencing (WGS) and Cas9 binding assays.[45]

WGS approaches hold the greatest promise for unambiguous characterization of editing effects throughout the genome. In theory, these methods should identify mutations occurring throughout an edited genome, but they currently have restricted reliability because of incomplete sequence coverage and alignments, statistical uncertainties in genome assembly techniques for families of small-size reads, and the challenges of dealing with repetitive regions.[46] Additionally, the use of WGS is complicated by further uncertainties regarding comparability of test and reference genomes, the occurrence of single-nucleotide variations, and the limits of sensitivity to detect low-level mutations with shallow sequencing depths.[45] Improvements in bioinformatics and the increasing availability of reliable genome maps are helping WGS to achieve robustness at costs and analytical timespans that allow for routine use of WGS as a characterization tool.[47,48]

Additional approaches for assessing off-target mutations, including Cas9 binding assays and methodologies based on analysis for DSBs, have been developed for prokaryotic and mammalian systems but as of yet have not been extended to plant systems.[49–54]

2.4 Omics Characterization

In addition to genomics, a host of omic tools are being developed with potential use in characterization of gene-editing outcomes. Studies considering and comparing outcomes of omics research are rapidly increasing,[55] as are the sophistication of approaches used.[56] To date transcriptomic, metabolomic, epigenomic, and phenomic methodologies for plant characterization have not shown unambiguous changes attributable to a given genetic transformation (other than for the gene and trait being introduced), since they are within the range of natural variation. As such omics methods alone do not indicate a safety concern in regard to transgenic crops.[57] Since at this point in time omics outcomes do not lend themselves to clear interpretation, a tiered approach to discriminate adverse indicators from currently uninterpretable outcomes will be necessary. Considerable development and sharing of data will be needed before omics outcomes in and of themselves are directly relevant to regulatory assessment of gene-edited crops.[57]

2.5 Bioinformatics and Computational Characterization

Genome editing is facilitated by computational design tools that identify both target and off-target sites. Lack of knowledge regarding the reliability in these predictive tools complicates research and encumbers regulatory assessments. For implementations of CRISPR–Cas9 editing, bioinformatics tools are used to design gRNAs with optimized sequence length and nucleotide composition for highly specific binding site recognition within the target gene.[58] Most current bioinformatics predictions have not been robustly validated as to target specificity so there is uncertainty whether the predictions are accurate.[17] Therefore, more detailed follow-on studies are necessary to determine the frequencies of off-target mismatches that may have occurred in the genome-editing process.[45] Improved and validated predictive algorithms for determining off-target mismatch frequencies for genome-edited plants are needed. While progress in this area is being made for human therapeutic applications,[27] findings sometimes do not directly translate to plants where there is widespread genomic duplication for large-scale segments as well as at the ploidy level.

Plant genomics information is used not only for reagent design and prediction of targeting but also for genome characterization with NGS techniques. Current plant reference genomes often are of draft quality with missing regions or poor assembly.[59] Additionally, only a few lines or cultivars have been well sequenced for any given plant species and many crop plants lack an assembled genome.[60] Duplicate genes and gene families may be poorly represented for plant species that do have a reference-quality genome, which reduces their usefulness for design and characterization of genome edits. These limitations will be largely addressed as the quality and breadth of plant genomic information increases.

3. GOVERNANCE AND REGULATION OF PLANT GENOME EDITING

Depth of knowledge has a significant impact on how individuals and various segments of society perceive risk of a given technology.[61] The presence of regulatory regimes and standards that are specific to products of biotechnology suggests this technology has special societal significance.[31] Therefore, it is incumbent on the community of researchers engaged in plant genome editing to recognize that both regulators and civil society may posit questions regarding genome editing that may be deemed nonrelevant or insignificant to experts in the field. Anticipating questions and addressing public concern will be critical for enabling genome editing as a process for crop improvement. The plant science community will encounter these questions in terms of regulatory processes for new products, governance of their research activities, and in participatory processes that bring together interested and affected parties.

3.1 Regulation

Enabling the widespread development and use of crops improved through genome editing will require further clarification of regulatory standards for genetically modified (GM) crops and their applicability to genome-edited crops. Moreover, there must be resolution surrounding the current bipolar regulatory standards that invoke both the product and process of bioengineering in regulatory decisions and, thus, lead to confusion among regulators, product developers, and civil society regarding the extent and nature of concerns associated with genome-edited crops.

3.1.1 The View From Europe

The regulatory framework for GM crops in the European Union is well described by Sprink and colleagues.[62] In EU legislation, recombinant

DNA (rDNA) technology invokes regulation, while mutagenesis is explicitly excluded.[63] The definition of genetically modified organism (GMO) has proven malleable in terms of the divergent perspectives on regulation of genome-edited crops within the EU, since the statutory language can be interpreted as relating to the use of rDNA technology as a process to derive a product or to the occurrence of rDNA in the derived product.[62] Furthermore, the EU adherence to a precautionary principle enables the perception of incomplete scientific knowledge about risk to prohibit marketing of products considered to be hazardous (http://eur-lex.europa.eu/legal-content/EN/TXT/?uri=URISERV:l32042). This ponderous and open-ended deliberative process has disabled decision making for GMOs in the EU[64,65] and promises to do the same for genome-edited crops.[62]

The potential use of genome editing as a tool for crop improvement has been the subject of intense and prolonged deliberation within research and regulatory communities.[40,42,66–74] This has especially been the case throughout Europe where the initial expectation was that use of genome editing would allow bioengineering applications for crop improvement which are not possible using transgenic techniques (i.e., for GMOs) due to public objections.[68] This early optimism has been frustrated by extensive back and forth among interested and affected parties regarding the regulatory interpretation of genome-edited crops. Arguments for acceptance of gene-editing technology within the EU were at first based on extensive discussion among regulators, scientists, and policy planners who found little reason for regulation of genome-edited crops when only simple insertions/deletions were involved through site-directed mutation.[62] This initial viewpoint has been supplanted by increasing uncertainty as evidenced by the prolonged and currently unresolved proposed commercial introduction of herbicide-resistant canola created by genome editing using OMM. The timeline of regulatory and legal actions surrounding the EU debate for this product involves principally the subtleties of the EU definition of a GMO, which has language that can support various interpretations as to whether the intent is to regulate on the basis of the process used or the product developed.[62] This gives testament to the hurdles this technology will face throughout the EU, and while it brings into question whether the on-going product or process debate skirts the need to better engage broad-based public views toward new technologies,[75] it appears here and elsewhere there is a need to move toward a more pragmatic and scientifically defensible position based on product in terms of the regulatory actions that are ultimately made.[76]

Regulators within Europe have for the most part remained mute as to the ultimate regulatory fate of genome-edited crops. Certain states within Europe, however, have made their initial positions known. Swedish regulators currently view SDN1-derived plants[72] (Box 2) as not subject to regulation as long as the nuclease-encoding rDNA is absent due to NS selection.[62,77] A somewhat similar opinion has been voiced in Germany as well.[78,79]

3.1.2 Regulatory Status of Genome-Edited Crops in the United States

To date, the situation has been markedly different in the United States where the regulation of products of biotechnology occurs under the Coordinated Framework and uses existing regulation to address rDNA organisms.[80] The principal agencies working within the Coordinated Framework to evaluate bioengineered crops are the US Department of Agriculture (USDA), Environmental Protection Agency (EPA), and Food and Drug Administration (FDA), which address, respectively, effects to agriculture, the environment, and to human health. The statutes and processes for regulatory-compliant risk assessments for bioengineered crops in the United States are well described elsewhere[81]; and although these regulations are commonly considered product focused, they in fact involve a de facto process-based trigger. Regulation by USDA, for example, is largely predicated on introduction of rDNA using *Agrobacterium tumefaciens* through regulatory language developed from USDA statutory remit under the Plant Protection Act regarding plant pests and noxious weeds.

The USDA Animal and Plant Health Inspection Service (APHIS) has been the most visible US agency in addressing genome-edited crops. Since 2004, APHIS has seen a steady stream of inquiries regarding genome-edited crops through the "Am I Regulated?" portal where it responds to Regulated Articles Letters of Inquiry regarding the regulatory status of proposed products (https://www.aphis.usda.gov/aphis/ourfocus/biotechnology/am-i-regulated). Responses to date to these inquires indicate that the USDA does not view genome-edited crops involving simple insertion/deletions, with limited numbers of bases and the absence of transgenic elements in the finished product, as subject to regulation (SDN1-derived products, Box 2).[76] Instances involving small template or directed transgene insertions would appear to remain of regulatory interest[82]; and in a least one instance, *A. tumefaciens* introduction of rDNA was not of regulatory concern when the subsequent removal of transgenic elements was demonstrated

BOX 2 Genome-Editing Approaches and Categories

In case-by-case assessments of biotechnology-derived plant products, regulators initially seek understanding as to the degree the methodology employed is familiar, as this helps to inform assessment of risks and uncertainties that may be associated with a given trait within the crop of interest. In terms of genome-edited crops, regardless of the reagent system used, an understanding of potential downstream outcomes for the plant product can be ascertained through consideration of the editing approach and the introduction and fate of the reagent.[76]

Accomplishing a genome edit uses a site-directed nuclease (SDN) to cause DSBs. Repair of the DSBs occurs by various mechanisms: NHEJ which randomly insert/deletes one to several bases to cause point mutation (SDN1) and homology-directed repair (HDR) involving native or synthetic template insertion (SDN2) or transgene insertion (SDN3). These approaches represent, respectively, insertion of progressively consequential nucleotide base sequences at a targeted DSB. Approaches involving OMM involve simple nucleotide replacements and are analogous to SDN1 to the degree the edits are viewed as point mutations and not template insertions.[74,121]

Additionally, in accomplishing the genome edit, the technique for introduction of the GEEN may be transient in the form of the protein, protein/RNA complex (e.g., Cas9/gRNA ribonucleoprotein complex), mRNA, or DNA that does not integrate into the host genome (category 1); may introduce nuclease-encoding rDNA in the genome that is subsequently removed through NS selection (category 2); or may involve stable genome integration of nuclease-encoding rDNA (category 3). These categories represent a progression of increasingly consequential procedures ranging from transient insertion of short-lived ribonucleoprotein complexes into the cytoplasm of cells to the introduction of nuclease-encoding rDNA into the host genome.

Together, the approach to gene editing used and the category describing introduction and fate of the nuclease roughly reflect the degree of regulatory uncertainty regarding the derived phenotype. The topology of current regulatory views concerning a given outcome of genome editing in terms of these factors are shown below (as adapted from Wolt and colleagues[76]). An additional layer of regulatory consideration with respect to category 2 and category 3 will be whether the design of the reagent has the potential to enable a gene drive.

	Category 1	Category 2	Category 3
SDN1	NHEJ or nucleotide replacement with transient introduction of reagent	NHEJ with stable integration of nuclease-encoding rDNA and subsequent NS selection	NHEJ with stable integration of nuclease-encoding rDNA
SDN2	HDR with transient introduction of reagent	HDR with stable integration of nuclease-encoding rDNA and subsequent NS selection	HDR with stable integration of nuclease-encoding rDNA
SDN3	Transient introduction of reagent with site-directed transgene insertion	Site-directed transgene insertion with stable integration of nuclease-encoding rDNA and NS selection	Site-directed transgene insertion with stable integration of nuclease-encoding rDNA

(https://www.aphis.usda.gov/biotechnology/downloads/reg_loi/aphis_resp_isu_ting_rice.pdf). Thus, a one-size-fits-all approach cannot be applied to USDA regulatory opinion regarding genome-edited crops. While there are no clear opinions of genome-edited crops from within EPA and FDA, the operational paradigm of agencies working through the Coordinated Framework is consensus based and so, it should be anticipated these agencies will hold similar current perspectives as seen at USDA.

In recognition of the rapid pace of biotechnology innovation and its importance to emergence of the bioeconomy, the US Executive Office of the President in 2015 directed FDA, EPA, and USDA to develop an update to the Coordinated Framework. In addition, these agencies were tasked to work through an interagency Biotechnology Working Group to formulate a long-term approach to future products of biotechnology and to commission an external, independent analysis to help inform future policy in this area.[83] In parallel with this activity, APHIS initiated a regulatory rulemaking process that proposed defining products of biotechnology as subject to regulation with explicit exclusion of conventionally and mutagenically derived organisms.[84] The distinction of product vs process as a regulatory trigger is complicated through a suggested definition of products of biotechnology that considers the process involved in their development. By this definition, processes involving insertion or deletion of DNA segments and directed genome altering (i.e., genome editing) could be subject to future regulation within the United States.

In early 2017, both USDA and FDA announced further actions to update regulatory oversight for biotechnology with implications to future development of genome-edited crops. The extensive APHIS-proposed revisions to its biotechnology regulations (82 FR 7008, 1/19/2017) include exclusion from regulatory oversight for genetic modifications that are solely a deletion—regardless of size—or a single base pair substitution that could otherwise result from the use of chemical- or radiation-based mutagenesis. Additionally, NS progeny are excluded from regulation for a bioengineered organism where the donor nucleic acid inserted into the recipient genome is not passed to the recipient progeny and has not altered the DNA sequence of the progeny. These exclusions are highly restrictive and, if adopted, will leave the vast majority of genome-edited plants subject to some form of regulatory oversight by APHIS. At the same time, FDA has requested public comment as to the use of genome editing for creation of plant varieties used for food and feed (82 FR 6564, 1/19/2017). The interest by FDA is to determine if genome editing represents unique categories of plant varieties for which food risks may need consideration. Together these actions portend continuing and protracted uncertainty as to the ultimate regulatory oversight which may be extended to genome-edited crops in the United States.

3.1.3 Regulatory Opinion Elsewhere

In Canada, regulation is based on plants with novel traits (PNTs) which broadly encompasses plants modified by genetic engineering or mutagenesis or plants without a history of production and safe consumption in Canada.[85] Genome-edited crops fit readily within the context of PNTs and therefore are expected to be accommodated by existing Canadian regulation. Herbicide tolerant canola (*Brassica napus*) developed through the use of OMM has been approved in Canada as a PNT.[76]

In 2012 and 2013, Food Standards Australia New Zealand convened expert scientific panels to consider regulation of different plant breeding techniques.[86] With respect to use of OMM or SDN1 for genome editing, the consensus was that these should not be considered GM, whereas SDN2 and SDN3 techniques should[87] (Box 2). The Australian Office of the Gene Technology Regulator has largely adopted this standard,[88] whereas legal opinion within New Zealand determined this was a legislative rather than regulatory matter, thus genome-editing technologies currently stand as GM and subject to regulation.[89]

In Argentina, the case-by-case determination of new breeding technology, including genome editing, is considered under the Law on the Promotion of the Development and Production of Modern Biotechnology[62] and is focused on the intended traits rather than the process used to generate the trait. Although the framework whereby a genome-edited trait will be evaluated has been elaborated by Argentine authorities,[87] specific instances have not yet been reported.

Elsewhere throughout the world, the lack of precedent regarding crops developed through genome editing leaves no clear picture as to regulatory opinion. While early sentiment of regulators from numerous countries and regions has tended toward viewing use of SDN1 or OMM as instances of mutagenesis and not subject to regulation,[68] the rapid pace of GEEN applications and the complex continuum of processes extending from simple insertions/deletions to HDR with native template or transgene insertions (SDN2 and SDN3) will result in case-by-case consideration as an important determinant as to how regulatory standards are ultimately resolved. Therefore, it will be important for the plant bioengineering community—both public and commercial enterprise—to push forward relevant cases for consideration within various regulatory domains.

3.1.4 Regulatory Trends for Genome-Edited Crops

Widespread confusion on the part of all parties concerned with development and regulation of genome-edited crops comes from the tendency toward oversimplification in descriptions of a process that involves a continuum of techniques extending from simple site-specific mutation to site-directed insertions of transgenes (Box 2). While there is a tendency by regulators to favor no regulation of SDN1-derived crops, products involving other GEEN approaches may be of regulatory interest, including methods that do not involve removal of nuclease-encoding rDNA elements and SDN2 and SDN3 methods that involve novel template or transgene insertions, respectively.[41,76] Furthermore, multiplexing to generate simultaneous gene edits will likely be of regulatory interest because of the possibility for rearrangements.[27] And clearly the enabling of gene drives through genome editing will be of significance to regulators until issues surrounding the environmental release of gene drives have been clarified.[34] The EU precautionary approach as applied to GMOs is increasingly reflected elsewhere in the world where new biosafety laws are being implemented[90–92] and is likely to have a substantive effect on international regulatory opinion regarding genome-edited crops.[76]

3.2 Research Governance

The biotechnology research community has a well-established history of governance to assure safety of emerging processes beginning with the Asilomar Conference,[93] which has important ramifications for building public trust and for informing the nature of regulatory standards applied to products of bioengineering. This responsibility for research governance is evidenced in genome engineering where the research community has considered appropriate needs for upstream computational prediction and downstream toxicity testing for therapeutics developed through genome editing,[27] and has given particular emphasis to environmental considerations relating to design and deployment of gene drives.[35,94] Outside of mainstream research, do-it-yourself biology (DIYbio) communities in both

Europe and the United States have established governance mechanisms and codes of conduct which include safety and security.[95,96] Governance mechanisms of this sort will become increasingly important to the plant science community as the pace and accessibility of plant bioengineering techniques and derived products accelerates.

Research involving recombinant or synthetic nucleic acid molecules that is funded by the National Institutes of Health at institutions in the United States and abroad follows the *Guidelines for Research Involving Recombinant or Synthetic Nucleic Acid Molecules (NIH Guidelines)* (http://osp.od.nih.gov/office-biotechnology-activities/biosafety/nih-guidelines). This guidance establishes safety practices and containment procedures to protect researchers, the public, and the environment and represents a strong system of governance within the research community. Governance is achieved through the oversight of Institutional Biosafety Committees (IBCs) whose remit for oversight includes material transfers among laboratories and field releases of noncommercialized genetically engineered organisms that are subject to APHIS permitting or notification procedures. In the United States, other federal agencies (e.g., the Departments of Energy, Veterans Affairs, and Agriculture) that fund or conduct research with recombinant or synthetic nucleic acid require compliance with NIH Guidelines. Apart from federal guidelines, several IBCs at public institutions have extended governance to CRISPR technology and the design of gene drive experiments (http://www.compliance.iastate.edu/committees/ibc/forms) and may require containment plans for research with gene drives (http://blink.ucsd.edu/sponsor/EHS/forms-ehs/containment-plan-gene-drives.html).

The NIH Guidelines represent a tool for guiding the research community to recognize the potential biosafety implications of recombinant or synthetic nucleic acid research and obligations of researchers in terms of biosafety and biosecurity. Although not all government and privately funded research is governed by these guidelines, the impact of NIH Guidelines is far reaching. Additional agencies of government in recognition of the safety and security implications of products produced through bioengineering have undertaken outreach to build knowledge and encourage governance structures for genome-editing research within diverse communities.[34,97,98]

With respect to gene drives, the National Academies[34] have recommended research governance to establish phased testing and other scientific approaches for reducing gene drive potential. Because of the dual-use potential of some CRISPRs, the Defense Advanced Research Projects

Agency is supporting research on strategies for appropriate design, deployment, and recall of genome-edited products (http://www.darpa.mil/news-events/2016-09-07).

3.3 Public Participation

As noted, both the conventional research community and the increasingly extensive DIYbio community are engaged in governance that helps to address safety and security of genome-editing technology, which provides direction to evolving regulatory and policy decisions surrounding the uses of genome editing. Broader civil society also makes increasingly strong claims to governance of emerging technology,[99,100] and this voice must be heard as it does not necessarily align with scientific opinion when products of modern biotechnology are considered.

The importance of public engagement was highlighted by recent studies in the United States which indicated fundamental differences between scientists and civil society regarding health and safety of GM foods. In a 2015 survey by the Pew Trust in cooperation with the American Association for the Advancement of Science (AAAS), 88% of AAAS scientists surveyed considered GM foods as safe to eat, whereas only 37% of US adults surveyed shared this sentiment.[101] Further, in a subsequent survey, 67% of adults felt scientists did not have a clear understanding of the health effects of GM crops.[102] And among US adults in this survey, 24% of those who care a great deal about the use of GM foods expressed a low level of trust in scientists to provide accurate information.

Simple one-way communication of scientific understanding to the public, which is commonly filtered through media reporting, is unlikely to positively influence growing public mistrust of science. As seen in other realms, the media paradigm of "balanced reporting" to voice opposing scientific views may perpetrate informational bias regarding scientific opinions.[103] When reporting fails to state the degree of scientific agreement that exists, the reporting of opposing scientific views strongly skews the public to mistrust science and leads to negative views of technology. Broad-based scientific understanding of an issue is often disregarded because skeptical views from scientific minorities can lead the general public to significantly doubt scientific consensus.[104] Establishing and conveying the degree of scientific consensus regarding an issue has a strong impact on public acceptance because of individual responsiveness to descriptive social fact (normative information).[105] Scientific mistrust can be countered to some extent through simple messages conveying the degree of scientific consensus, especially when numerically based and when individuals are first encouraged to express their estimate of the degree of scientific agreement that exists on an issue.[106]

This of course necessitates considerations of improved knowledge dissemination among scientists and civil society and is reflected in an increasing emphasis on dialog among widely varied social groups rather than one-way communication from technical experts.[100] While the arguments for participatory processes for technology assessment are strong,[99] the realization of workable means to accomplish effective decision making within an analytical-deliberative frame has yet to be demonstrated.[64] Issues of participant motivations, rules of discourse, group composition, achieving closure, and connection to other democratic processes are not yet well understood.[107]

Given the continuing erosion in public trust in scientific opinion,[100] especially with regard to the increasingly commercially dominated and strongly hyped fields of biotechnology,[108,109] it is important to recognize major concerns likely to be voiced by civil society in terms of CRISPR–Cas9 and its potential crop uses. These include the balancing of risk and benefits of the technology, the potential for ecological disequilibrium—in particular with respect to interspecies horizontal transfer of gene drives, and the ways intellectual property and regulation control how and by whom gene editing is used.[110] Until a workable scientific and public consensus is achieved, addressing real or perceived issues regarding the environmental impact and other concerns of bioengineered products could slow the adoption of genome-edited crops much as it has for GM crops.[64] Addressing these issues at formative stages of the technology, when funding policies and directions are being developed, will be preferred to later stages of product development when stances toward technology adoption may have hardened.[111]

4. SUMMARY AND CONCLUSIONS

4.1 Opportunities for Future Research and Stewardship of Genome-Edited Plants

Scientific clarity regarding the safety and security of genome editing for the development of useful traits in crops will be needed to build public trust and understanding of this technology as well as to inform the appropriate regulatory path for product commercialization. Key scientific needs in this area are improved methodologies to predict and confirm off-target effects within the genome. Furthermore, there is a need for greater understanding of the relationship of off-target effects within the genome to potential unanticipated downstream effects of genome-edited crops to ecosystems where deployed or to consumers of derived products. Addressing this will require further advances in genomic and other omics techniques and their interpretation, improved bioinformatics tools and models, and relevant information regarding integration of knowledge ranging from molecular to landscape scales.

Additionally, methodological advances in the process of genome editing may help to alleviate concerns regarding unanticipated direct and indirect effects. Direct nuclease introduction vs use of nuclease-encoding DNA can avoid concerns regarding long-term fate of editing reagents. Alternatively, use of self-limiting designs, such as daisy chains, can greatly reduce the potential for initiating genome-editing systems with gene drive potential. Dual-use considerations of genome editing in crops can be anticipated and addressed through continued development and improvement of existing physical and biological safeguards and with experimental protocols that take a staged approach to the development and deployment of gene-edited crops.

4.2 Policy for Investigation, Development, and Deployment of Gene-Edited Crops

Governance mechanisms for research can inform and reduce the burden for regulation of genome-edited crops. These would include product development considerations to assure safety and security—such as improved approaches for rational product design, genotypic and phenotypic characterization, evaluation, release, and monitoring that integrate safety and security considerations into the development process—and the communication of these approaches in a transparent manner. Governance of genome-editing technology should also consider inclusiveness and extend public engagement beyond simple expert communication to encompass mechanisms that more fully engage interested and affected parties. Engagement of the public is most likely to be impactful when it is done as technology emerges and will likely have greater effect when focused on science funding policy rather than when focused on regulatory processes that occur further downstream in the product development process.

Finally, the regulatory path forward for genome-edited crops is currently trapped in semantic and statutory arguments regarding existing regulatory processes for GMOs. These arguments must be overcome if there is to be a long-term solution for plant bioengineering, since the nature and direction of innovation will likely outpace regulatory concepts that are based on process considerations. Understanding risks and uncertainties associated with processes for genome engineering is critical as new technological approaches emerge; but once understanding builds within the science community and is subject to social discourse, the ultimate regulatory standards for bioengineered crops need to focus on the attributes of the crop phenotype and its derived products regardless of the process which may have been employed in development.

REFERENCES

1. Doudna JA, Charpentier E. The new frontier of genome engineering with CRISPR-Cas9. *Science*. 2014;346(6213):1258096.
2. Hsu PD, Lander ES, Zhang F. Development and applications of CRISPR-Cas9 for genome engineering. *Cell*. 2014;157(6):1262–1278.
3. Beetham PR, Kipp PB, Sawycky XL, Arntzen CJ, May GD. A tool for functional plant genomics: chimeric RNA/DNA oligonucleotides cause in vivo gene-specific mutations. *Proc Natl Acad Sci USA*. 1999;96(15):8774–8778.
4. Townsend JA, Wright DA, Winfrey RJ, et al. High-frequency modification of plant genes using engineered zinc-finger nucleases. *Nature*. 2009;459(7245):442–445.
5. Wolt JD, Wang K, Yang B. The regulatory status of genome-edited crops. *Plant Biotechnol J*. 2016;14(2):510–518.
6. Li T, Liu B, Spalding MH, Weeks DP, Yang B. High-efficiency TALEN-based gene editing produces disease-resistant rice. *Nat Biotechnol*. 2012;30(5):390–392.
7. Shukla VK, Doyon Y, Miller JC, et al. Precise genome modification in the crop species Zea mays using zinc-finger nucleases. *Nature*. 2009;459(7245):437–441.
8. Belhaj K, Chaparro-Garcia A, Kamoun S, Patron NJ, Nekrasov V. Editing plant genomes with CRISPR/Cas9. *Curr Opin Biotechnol*. 2015;32:76–84.
9. Schiml S, Fauser F, Puchta H. The CRISPR/Cas system can be used as nuclease for in planta gene targeting and as paired nickases for directed mutagenesis in Arabidopsis resulting in heritable progeny. *Plant J*. 2014;80(6):1139–1150.
10. Gaj T, Gersbach CA, Barbas CF. ZFN, TALEN, and CRISPR/Cas-based methods for genome engineering. *Trends Biotechnol*. 2013;31(7):397–405.
11. Osakabe Y, Osakabe K. Genome editing with engineered nucleases in plants. *Plant Cell Physiol*. 2015;56(3):389–400.
12. Kleinstiver BP, Prew MS, Tsai SQ, et al. Engineered CRISPR-Cas9 nucleases with altered PAM specificities. *Nature*. 2015;523(7561):481–485.
13. Kleinstiver BP, Pattanayak V, Prew MS, et al. High-fidelity CRISPR-Cas9 nucleases with no detectable genome-wide off-target effects. *Nature*. 2016;529(7587):490–495.
14. Slaymaker IM, Gao L, Zetsche B, Scott DA, Yan WX, Zhang F. Rationally engineered Cas9 nucleases with improved specificity. *Science*. 2016;351(6268):84–88.
15. Hsu PD, Scott DA, Weinstein JA, et al. DNA targeting specificity of RNA-guided Cas9 nucleases. *Nat Biotechnol*. 2013;31(9):827–832.

16. Pattanayak V, Lin S, Guilinger JP, Ma E, Doudna JA, Liu DR. High-throughput profiling of off-target DNA cleavage reveals RNA-programmed Cas9 nuclease specificity. *Nat Biotechnol.* 2013;31(9):839–843.
17. Wolt JD, Wang K, Sashital D, Lawrence-Dill CJ. Achieving plant CRISPR targeting that limits off-target effects. *Plant Genome.* 2016;9(3). http://dx.doi.org/10.3835/plantgenome2016.05.0047.
18. Lowder L, Malzahn A, Qi Y. Rapid evolution of manifold CRISPR systems for plant genome editing. *Front Plant Sci.* 2016;7:1683.
19. Mao Y, Botella JR, Zhu J-K. Heritability of targeted gene modifications induced by plant-optimized CRISPR systems. *Cell Mol Life Sci.* 2017;74(6):1075–1093.
20. Song G, Jia M, Chen K, et al. CRISPR/Cas9: a powerful tool for crop genome editing. *Crop J.* 2016;4(2):75–82.
21. Jacobs TB, LaFayette PR, Schmitz RJ, Parrott WA. Targeted genome modifications in soybean with CRISPR/Cas9. *BMC Biotechnol.* 2015;15(1):1.
22. Sun X, Hu Z, Chen R, et al. Targeted mutagenesis in soybean using the CRISPR-Cas9 system. *Sci Rep.* 2015;5:10342.
23. Curtin SJ, Zhang F, Sander JD, et al. Targeted mutagenesis of duplicated genes in soybean with zinc-finger nucleases. *Plant Physiol.* 2011;156(2):466–473.
24. Nemudryi A, Valetdinova K, Medvedev S, Zakian S. TALEN and CRISPR/Cas genome editing systems: tools of discovery. *Acta Naturae.* 2014;6(3):19–40 (англоязычная версия).
25. Chu VT, Weber T, Wefers B, et al. Increasing the efficiency of homology-directed repair for CRISPR-Cas9-induced precise gene editing in mammalian cells. *Nat Biotechnol.* 2015;33(5):543–548.
26. Renaud J-B, Boix C, Charpentier M, et al. Improved genome editing efficiency and flexibility using modified oligonucleotides with TALEN and CRISPR-Cas9 nucleases. *Cell Rep.* 2016;14(9):2263–2272.
27. Corrigan-Curay J, O'Reilly M, Kohn DB, et al. Genome editing technologies: defining a path to clinic. *Mol Ther.* 2015;23(5):796.
28. Curtin SJ, Voytas DF, Stupar RM. Genome engineering of crops with designer nucleases. *Plant Genome.* 2012;5(2):42–50.
29. Martin-Ortigosa S, Peterson DJ, Valenstein JS, et al. Mesoporous silica nanoparticle-mediated intracellular cre protein delivery for maize genome editing via loxP site excision. *Plant Physiol.* 2014;164(2):537–547.
30. Svitashev S, Schwartz C, Lenderts B, Young JK, Cigan AM. Genome editing in maize directed by CRISPR–Cas9 ribonucleoprotein complexes. *Nat Commun.* 2016;7:13274.
31. Nuffield Council on Bioethics. *Genome Editing. An Ethical Review.* London: Nuffield Council on Bioethics; 2016.
32. Araki M, Ishii T. Towards social acceptance of plant breeding by genome editing. *Trends Plant Sci.* 2015;20(3):145–149.
33. Anderson JE, Michno J-M, Kono TJ, et al. Genomic variation and DNA repair associated with soybean transgenesis: a comparison to cultivars and mutagenized plants. *BMC Biotechnol.* 2016;16(1):1.
34. National Academies of Sciences, Engineering, and Medicine. *Gene Drives on the Horizon: Advancing Science, Navigating Uncertainty, and Aligning Research with Public Values.* Washington, DC: The National Academies Press; 2016.
35. Akbari OS, Bellen HJ, Bier E, et al. Safeguarding gene drive experiments in the laboratory. *Science.* 2015;349(6251):927–929.
36. Esvelt KM, Smidler AL, Catteruccia F, Church GM. Concerning RNA-guided gene drives for the alteration of wild populations. *Elife.* 2014;3:e03401.

37. DiCarlo JE, Chavez A, Dietz SL, Esvelt KM, Church GM. Safeguarding CRISPR-Cas9 gene drives in yeast. *Nat Biotechnol.* 2015;33:1250–1255.
38. Noble C, Min J, Olejarz J, et al. Daisy-chain gene drives for the alteration of local populations. *bioRxiv.* 2016;057307. http://dx.doi.org/10.1101/057307.
39. Jones HD. Regulatory uncertainty over genome editing. *Nat Plants.* 2015;1:14011.
40. Pauwels K, Podevin N, Breyer D, Carroll D, Herman P. Engineering nucleases for gene targeting: safety and regulatory considerations. *N Biotechnol.* 2014;31(1):18–27.
41. Wolt JD, Yang B, Wang K, Spalding MH. Regulatory aspects of genome-edited crops. *In Vitro Cell Dev Biol Plant.* 2016;52(4):349–353.
42. European Food Safety Authority Panel on Genetically Modified Organisms. Scientific opinion addressing the safety assessment of plants developed using zinc finger nuclease 3 and other site-directed nucleases with similar function. *EFSA J.* 2012;10:2943.
43. Schnell J, Steele M, Bean J, et al. A comparative analysis of insertional effects in genetically engineered plants: considerations for pre-market assessments. *Transgenic Res.* 2015;24(1):1–17.
44. Zischewski J, Fischer R, Bortesi L. Detection of on-target and off-target mutations generated by CRISPR/Cas9 and other sequence-specific nucleases. *Biotechnol Adv.* 2017;35:95–104.
45. O'Geen H, Abigail SY, Segal DJ. How specific is CRISPR/Cas9 really? *Curr Opin Chem Biol.* 2015;29:72–78.
46. Pauwels K, De Keersmaecker SC, De Schrijver A, du Jardin P, Roosens NH, Herman P. Next-generation sequencing as a tool for the molecular characterisation and risk assessment of genetically modified plants: added value or not? *Trends Food Sci Technol.* 2015;45(2):319–326.
47. Ekblom R, Wolf JB. A field guide to whole-genome sequencing, assembly and annotation. *Evol Appl.* 2014;7(9):1026–1042.
48. Guttikonda SK, Marri P, Mammadov J, et al. Molecular characterization of transgenic events using next generation sequencing approach. *PLoS One.* 2016;11(2):e0149515.
49. Mali P, Aach J, Stranges PB, et al. CAS9 transcriptional activators for target specificity screening and paired nickases for cooperative genome engineering. *Nat Biotechnol.* 2013;31(9):833–838.
50. Wu X, Scott DA, Kriz AJ, et al. Genome-wide binding of the CRISPR endonuclease Cas9 in mammalian cells. *Nat Biotechnol.* 2014;32(7):670–676.
51. Tsai SQ, Zheng Z, Nguyen NT, et al. GUIDE-seq enables genome-wide profiling of off-target cleavage by CRISPR-Cas nucleases. *Nat Biotechnol.* 2015;33(2):187–197.
52. Frock RL, Hu J, Meyers RM, Ho Y-J, Kii E, Alt FW. Genome-wide detection of DNA double-stranded breaks induced by engineered nucleases. *Nat Biotechnol.* 2015;33(2):179–186.
53. Kim D, Bae S, Park J, et al. Digenome-seq: genome-wide profiling of CRISPR-Cas9 off-target effects in human cells. *Nat Methods.* 2015;12(3):237–243.
54. Yan WX, Mirzazadeh R, Garnerone S, et al. BLISS: quantitative and versatile genome-wide profiling of DNA breaks in situ. *bioRxiv.* 2016;091629. http://dx.doi.org/10.1101/091629.
55. Ricroch AE. Assessment of GE food safety using '-omics' techniques and long-term animal feeding studies. *N Biotechnol.* 2013;30(4):349–354.
56. Ibáñez C, Simó C, García-Cañas V, Acunha T, Cifuentes A. The role of direct high-resolution mass spectrometry in foodomics. *Anal Bioanal Chem.* 2015;407(21):6275–6287.
57. National Academies of Sciences, Engineering, and Medicine. *Genetically Engineered Crops: Experiences and Prospects.* Washington, DC: The National Academies Press; 2016. http://dx.doi.org/10.17226/23395.
58. Brazelton Jr VA, Zarecor S, Wright DA, et al. A quick guide to CRISPR sgRNA design tools. *GM Crops Food.* 2015;6(4):266–276.

59. Tello-Ruiz MK, Stein J, Wei S, Youens-Clark K, Jaiswal P, Ware D. Gramene: a resource for comparative analysis of plants genomes and pathways. In: Edwards D, ed. *Plant Bioinformatics: Methods and Protocols, Methods in Molecular Biology*. 2016; vol. 1374:141–163. http://dx.doi.org/10.1007/978-1-4939-3167-5_7.
60. Shangguan L, Han J, Kayesh E, et al. Evaluation of genome sequencing quality in selected plant species using expressed sequence tags. *PLoS One*. 2013;8(7):e69890.
61. Science Communication Unit. Science for environmental policy future brief: public risk perception and environmental policy. Report Produced for the European Commission DG Environment, September 2014. Bristol: University of the West of England; 2014.
62. Sprink T, Eriksson D, Schiemann J, Hartung F. Regulatory hurdles for genome editing: process-vs. product-based approaches in different regulatory contexts. *Plant Cell Rep*. 2016;35(7):1493–1506.
63. European Parliament. Directive 2001/18/EC of the European Parliament and of the Council of 12 March 2001 on the deliberate release into the environment of genetically modified organisms and repealing Council Directive 90/220/EEC. *Off J Eur Community*. 2001;L 106:1–38.
64. Smyth SJ, McDonald J, Falck-Zepeda J. Investment, regulation, and uncertainty: managing new plant breeding techniques. *GM Crops Food*. 2014;5(1):44–57.
65. Smyth SJ, Phillips PW. Risk, regulation and biotechnology: the case of GM crops. *GM Crops Food*. 2014;5(3):170–177.
66. Gruère GP, Rao S. A review of international labeling policies of genetically modified food to evaluate India's proposed rule. *AgBioForum*. 2007;10(1):51–64.
67. Breyer D, Herman P, Brandenburger A, et al. Commentary: genetic modification through oligonucleotide-mediated mutagenesis. A GMO regulatory challenge? *Environ Biosaf Res*. 2009;8(2):57–64.
68. Lusser M, Parisi C, Plan D, Rodriguez-Cerezo E. *New Plant Breeding Techniques: State-of-the-Art and Prospects for Commercial Development*. Brussels, Belgium: European Commission Joint Research Centre; 2011.
69. Lusser M, Rodríguez-Cerezo E. Comparative regulatory approaches for new plant breeding techniques. Paper Presented At: Workshop Proceedings. European Commission. JRC Technical Report EUR 2012.
70. Lusser M, Parisi C, Plan D, Rodríguez-Cerezo E. Deployment of new biotechnologies in plant breeding. *Nat Biotechnol*. 2012;30(3):231–239.
71. Podevin N, Devos Y, Davies HV, Nielsen KM. Transgenic or not? No simple answer! *EMBO Rep*. 2012;13(12):1057–1061.
72. Podevin N, Davies HV, Hartung F, Nogue F, Casacuberta JM. Site-directed nucleases: a paradigm shift in predictable, knowledge-based plant breeding. *Trends Biotechnol*. 2013;31(6):375–383.
73. International Life Sciences Institute. ILSI Workshop on New Breeding Technologies (NBT). Canaberra: International Life Sciences Institute SEA Region Australasia; June 19, 2013 2013.
74. Lusser M, Davies HV. Comparative regulatory approaches for groups of new plant breeding techniques. *N Biotechnol*. 2013;30(5):437–446.
75. Kuzma J. Policy: reboot the debate on genetic engineering. *Nature*. 2016;531: 165–167.
76. Wolt JD, Wang K, Yang B. The regulatory status of genome-edited crops. *Plant Biotechnol J*. 2016;14(2):510–518.
77. Swedish Board of Agriculture. *Is Regulatory Oversight Needed for Field Trials of Plants Produced Using Genome Editing Technology?* Jönköping: Jordbruksverket; 2015.
78. German Biosafety Commission (ZKBS). *Position Statement of the ZKBS on New Plant Breeding Techniques*. Berlin: German Federal Office of Consumer Protection and Food Safety; 2012.

79. German Federal Office of Consumer Protection and Food Safety (BVL). *Opinion on the Legal Classification of New Plant Breeding Techniques, in Particular ODM and CRISPR-Cas9*. Braunschweig: Bundesamt für Verbraucherschutz und Lebensmittelsicherheit; 2015, 11.
80. Office of Science and Technology Policy. Coordinated framework for regulation of biotechnology. *Fed Regist*. 1986;51:23302–23350.
81. McHughen A, Smyth S. US regulatory system for genetically modified [genetically modified organism (GMO), rDNA or transgenic] crop cultivars. *Plant Biotechnol J*. 2008;6(1):2–12.
82. Camacho A, Van Deynze A, Chi-Ham C, Bennett AB. Genetically engineered crops that fly under the US regulatory radar. *Nat Biotechnol*. 2014;32(11):1087–1091.
83. Executive Office of the President. Memorandum for Heads of Food and Drug Administration, Environmental Protection Agency, and Department of Agriculture: Modernizing the Regulatory System for Biotechnology Products, July 2, 2015. Washington, DC, 2015.
84. Animal and Plant Health Inspection Service. Environmental impact statement: introduction to the products of biotechnology. *Fed Regist*. 2016;81(24):6225–6229.
85. Smyth S, McHughen A. Regulating innovative crop technologies in Canada: the case of regulating genetically modified crops. *Plant Biotechnol J*. 2008;6(3):213–225.
86. Food Standards Australia New Zealand. New Plant Breeding Techniques, Workshop Report 2013. Wellington 2013.
87. Whelan AI, Lema MA. Regulatory framework for gene editing and other new breeding techniques (NBTs) in Argentina. *GM Crops Food*. 2015;6(4):253–265.
88. Schuttelaar and Partners. *The Regulatory Status of New Breeding Techniques in Countries Outside of the European Union*. The Hague, 2015.
89. Kershen DL. Sustainability Council of New Zealand Trust v. The Environmental Protection Authority: gene editing technologies and the law. *GM Crops Food*. 2015;6(4):216–222.
90. Bayer JC, Norton GW, Falck-Zepeda JB. Cost of compliance with biotechnology regulation in the Philippines: implications for developing countries. *AgBioForum*. 2010;13(1):53–62.
91. Gupta K, Karihaloo J, Khetarpal R. Biosafety regulations of Asia-Pacific countries. In: Asia-Pacific Association of Agricultural Research Institutions, Bangkok; Asia-Pacific Consortium on Agricultural Biotechnology, New Delhi and Food and Agricultural Organization of the United Nations, Rome; 2008:108.
92. Okeno JA, Wolt JD, Misra MK, Rodriguez L. Africa's inevitable walk to genetically modified (GM) crops: opportunities and challenges for commercialization. *N Biotechnol*. 2013;30(2):124–130.
93. Berg P, Baltimore D, Brenner S, Roblin RO, Singer MF. Summary statement of the Asilomar conference on recombinant DNA molecules. *Proc Natl Acad Sci USA*. 1975;72(6):1981–1984.
94. Oye KA, Esvelt K, Appleton E, et al. Regulating gene drives. *Science*. 2014; 345(6197):626–628.
95. Grushkin D, Kuiken T, Millet P. *Seven Myths and Realities About Do-It-Yourself Biology*. Washington, DC: Woodrow Wilson International Center for Scholars. 2013.
96. Kuiken T. *Our collective biology: enabling public science to build an ecosystem of makers in biology, June 1*. https://vimeo.com/album/3988159/video/169715336; 2016. Accessed January 1, 2017.
97. National Science Advisory Board for Biosecurity. *Strategies to Educate Amateur Biologists and Scientists in Non-Life Science Disciplines About Dual Use Research in the Life Sciences*. Washington, DC: National Science Advisory Board for Biosecurity; 2011.
98. Defense Advanced Research Projects Agency. Setting a safe course for gene editing research. http://www.darpa.mil/news-events/2016-09-07; 2016. Accessed December 26, 2016.

99. Van Oudheusden M. Where are the politics in responsible innovation? European governance, technology assessments, and beyond. *J Responsible Innov*. 2014;1(1):67–86
100. Durant J. Participatory technology assessment and the democratic model of the public understanding of science. *Sci Public Policy*. 1999;26(5):313–319.
101. Funk C, Rainie L. *Public and Scientists' Views on Science and Society*. Washington, DC: Pew Research Center; 2015.
102. Funk C, Kennedy B. *The New Fights: U.S. Public Divides Over Food Science*. Washington, DC: Pew Research Center; 2016.
103. Boykoff MT. Flogging a dead norm? Newspaper coverage of anthropogenic climate change in the United States and United Kingdom from 2003 to 2006. *Area*. 2007;39(4):470–481.
104. Aklin M, Urpelainen J. Perceptions of scientific dissent undermine public support for environmental policy. *Environ Sci Policy*. 2014;38:173–177.
105. van der Linden S, Lewandowsky S. *How to combat distrust of science*. https://www/scientificamerican.com/article/how-to-combat-distrust-of-science/; 2015. Accessed December 28, 2016.
106. Myers TA, Maibach E, Peters E, Leiserowitz A. Simple messages help set the record straight about scientific agreement on human-caused climate change: the results of two experiments. *PLoS One*. 2015;10(3):e0120985.
107. Carpini MXD, Cook FL, Jacobs LR. Public deliberation, discursive participation, and citizen engagement: a review of the empirical literature. *Ann Rev Polit Sci*. 2004;7:315–344.
108. Chalmers D, Nicol D. Commercialisation of biotechnology: public trust and research. *Int J Biotechnol*. 2004;6(2–3):116–133.
109. Master Z, Resnik DB. Hype and public trust in science. *Sci Eng Ethics*. 2013;19(2):321–335.
110. Rodriguez E. Ethical issues in genome editing using Crispr/Cas9 system. *J Clin Res Bioeth*. 2016;7:266.
111. Chapotin SM, Wolt JD. Genetically modified crops for the bioeconomy: meeting public and regulatory expectations. *Transgenic Res*. 2007;16(6):675–688.
112. Esvelt KM, Mali P, Braff JL, Moosburner M, Yaung SJ, Church GM. Orthogonal Cas9 proteins for RNA-guided gene regulation and editing. *Nat Methods*. 2013;10(11):1116–1121.
113. Zetsche B, Gootenberg JS, Abudayyeh OO, et al. Cpf1 is a single RNA-guided endonuclease of a class 2 CRISPR-Cas system. *Cell*. 2015;163(3):759–771.
114. Burstein D, Harrington LB, Strutt SC, et al. New CRISPR–Cas systems from uncultivated microbes. *Nature*. 2017;542:237–241.
115. Kleinstiver BP, Prew MS, Tsai SQ, et al. Engineered CRISPR-Cas9 nucleases with altered PAM specificities. *Nature*. 2015;523(7561):481–485.
116. Shen B, Zhang W, Zhang J, et al. Efficient genome modification by CRISPR-Cas9 nickase with minimal off-target effects. *Nat Methods*. 2014;11(4):399–402.
117. Sauer NJ, Narváez-Vásquez J, Mozoruk J, et al. Oligonucleotide-mediated genome editing provides precision and function to engineered nucleases and antibiotics in plants. *Plant Physiol*. 2016;170(4):1917–1928.
118. Gao H, Smith J, Yang M, et al. Heritable targeted mutagenesis in maize using a designed endonuclease. *Plant J*. 2010;61(1):176–187.
119. Bogdanove AJ, Voytas DF. TAL effectors: customizable proteins for DNA targeting. *Science*. 2011;333(6051):1843–1846.
120. Carroll D. Genome engineering with zinc-finger nucleases. *Genetics*. 2011;188(4):773–782.
121. Hartung F, Schiemann J. Precise plant breeding using new genome editing techniques: opportunities, safety and regulation in the EU. *Plant J*. 2014;78(5):742–752.

INDEX

Note: Page numbers followed by "*f*" indicate figures, "*t*" indicate tables, and "*b*" indicate boxes.

D

H

Kolibri 360 GmbH
Ferdinand-Jühlke-Straße 7
99095 Erfurt, Deutschland
produktsicherheit@kolibri360.de